ON THE WATER

Also by Guy de la Valdène

The Fragrance of Grass

Red Stag

For a Handful of Feathers

Making Game: An Essay on Woodcock

ON THE WATER

— A Fishing Memoir —

Guy de la Valdène

Guilford, Connecticut

*I would like to thank
Colleen Daily, Patrick Smith,
and Janice Goldklang for their help
and support*

An imprint of Rowman & Littlefield

Distributed by NATIONAL BOOK NETWORK

British Library Cataloguing-in-Publication Information available

Library of Congress Cataloging-in-Publication Data available.

ISBN 978-1-4930-0793-6

♾™ The paper used in this publication meets the minimum requirements
of American National Standard for Information Sciences—Permanence of
Paper for Printed Library Materials, ANSI/NISO Z39.48-1992.

For Gil Drake
and
Fanny Malone

And only the enlightened can recall their former lives;
for the rest of us, the memories of past existences are
but glints of light, twinges of longing, passing shadows . . .

—Peter Matthiessen, **The Snow Leopard**

CONTENTS

FOREWORD

In the interest of full disclosure, I confess to being a friend of Guy de la Valdène, and therefore not an objective source. We were introduced to each other by Jimmy Buffett, a character in his own right, who'd told me stories about Guy that sounded like strands from Jimmy's songs.

When we first met, on a fly-fishing trip, Guy was deceptively well behaved. Over time his inner rascal emerged, and I don't think anyone has made me laugh harder while cringing at the subject matter.

For quite a while I didn't know he was a writer, because he's uncommonly humble about it. Eventually he sent me a small book he'd finished about his quail farm, and it was a gem. A few years later, when I learned he was working on something new, I began nagging him for the manuscript.

On the Water arrived at my house not long ago, and from the first page I was carried away. Although Guy and I have lived very different lives, we've been addicted to some of the same destinations, from the Florida Keys to the Bahamian Out Islands to western Montana. Even his boyhood memories of fishing the moat at his family's French estate struck home, because the thrill with which he describes a pike's strike at twilight stirred my own recollections of stalking bass at dusk in the Everglades.

Of youthful summers in the Bahamas, Guy writes, "Imagine a mirror granting every wish, and then imagine looking into that mirror every single day." You could care nothing about fishing and still fall in love with that line.

In *On the Water*, Guy frames such magical reminiscences around a top-to-bottom account of how he engineered and vitalized a twenty-seven-acre pond on his farm near Tallahassee. I imagine his editor saying, "You want to write about *what*?"

"My little pond," Guy would have replied with the usual mischief in his smile.

He knew what to do. The book is a beauty.

His affection for nature's delicate details brings John McPhee to mind, and his awe of water's turbulent pull on the human soul makes me think of Norman Maclean. And then of course there's Thoreau, who was also famously inspired by the seasons of a pond.

A few years back I made a trip to Guy's farm. He put me in the bow of a small jon boat and gave me a tour of the water while I casted streamers at hungry largemouth bass. Naturally he knew every turtle, gator, moccasin, and osprey that appeared.

On the edge of the pond that Guy brought to life sits the small wooden house where he writes. It's baffling how he gets any work done—if it were me, I'd spend all day in front of the window watching for drama. The stagecraft of wild predation will never fail to make your heart thump.

Even if Guy wasn't my friend, even if I'd never laid eyes on the pond, I could honestly say that page after page of this book delivers glistening passages I wish I'd written myself. In such an admission there's no jealousy, only marvel.

Carl Hiaasen

PROLOGUE

Five years after the end of the Second World War, my sister and I boarded an ocean liner in New York City and sailed across the Atlantic to Cherbourg, France. A week later, after clearing customs and driving for what felt like an eternity, we were safely ensconced behind the walls of the estate our parents had purchased in Normandy. I was six years old. France was my new home.

Since none of the children in the adjoining village or any member of my parents' staff spoke English, my sister and I, surrounded by the rapid fire of the French dialect, learned how to speak our new language in six months. The following year, I was sent to boarding school thirty miles away. The long, gray winter months of Normandy—known as the chamber pot of France—were cold and dreary, but they were not as dreadful as some would make them out to be. There is a lot of fun to be had in boarding school, and I took advantage of it.

During the holidays, I spent much of my time exploring the waterways that crisscrossed the land I lived on. I would set traps in the mud burrows of water rats, build fish weirs across streams, and watch for high-floating feathers on the surface of ponds as a sign that the fall migration of ducks had begun. I dangled strips of red cloth from cane poles in the faces of frogs, gathered maggots for bait, tickled rainbow trout, studied the gait of long-legged birds that walked in a parody of purpose, and battled solitary pikes, emperors of their watery domains. I fished in creeks, swam in rivers, ate sandwiches on the shores of ponds, and once, when I was twelve years old, kissed a farm girl in the shadows of a chestnut tree next to a moat. When I was tired, I slept in the shadowy chambers of weeping willow trees and dreamed of the mischief I hoped to get into next.

Proximity to water has always been part of my life. On the land where I grew up there were two mapped rivers, a dozen streams and rivulets, subordinate farm ponds, and a moat that girdled the castle. Later it was on the saltwater flats of the Bahamas and then in the shallow waters of the Florida Keys where I learned to appreciate fish I could see and stalk, fish I could cast at, and fish that fought back. Once I knew how to use a fly rod, I added the Pacific Ocean and a number of rivers around the world to my angling settings. Now, as an older man with an aversion to crowded airports, a freshwater pond once again gathers my attention.

In 1990 I bought a farm in northern Florida that showcases red clay hills, live oak trees, and loblolly pines. There are doves in the fall, bobwhite quail in the winter, and turkeys in the spring. Sixty miles to the south there are redfish and flounder on the grass flats, sea trout and tarpon on the reef, and oysters on the beds. From the ten-acre spit of unproductive water that came with the property, I fashioned a twenty-seven-acre body of prime bass-fishing habitat. Its proximity to my house rekindles a number of childhood memories.

I share the pond with a great blue heron, a pair of Canada geese, bald eagles, nesting ospreys, anhingas, and quail year-round. The pond welcomes an avian community that migrates north and south to and from northern Florida, following seasonal and geological inclinations. The same characters or their offspring who visit the pond every year squabble over food, territory, sex, and property, each species behaving singularly like humans on a parallel migration from childhood to old age.

The water of the pond falls into dimness a foot below the surface, encouraging schools of largemouth bass, speckle perch, catfish, and bream to chase shad, shiners, and salamanders with the same diligence that pike, brown trout, and perch hunted roach, worms, and nymphs in France sixty years ago. Since that time, I have fished rivers and bays, and blue water and shallow flats, mostly with a fly rod and mostly in salt water. As a young man those were pursuits second only to chasing girls. These days my interest in fishing is more of a flirt than a passion. I love it all, but from a certain distance, a distance that betrays the intensity I deployed earlier in my life. It is more and more difficult to lure me from the comfort of my home beat.

In a small cabin with a porch and a wooden dock that extends thirty feet into the pond, I write almost every day at a desk that looks out over the water. Of late I have been comparing the seasons of the pond to the seasons of my life. Excluding the dog days of August, when the heat draws the pleasure out of rocking, I sit in a basket-weave chair on the deck and study the pond's activities just as someone more ambitious might watch the rise and fall of the stock market. In the evening I have a drink on the dock and, like the frogs and alligators, water snakes, bats, and other critters, shake off the intensity of the sun and consider my choices, encouraged as I am by the cooling influence of the surrounding water.

The mood of water adjusts itself to the change of seasons, the turn of the moon, the weather, and, perhaps most important, the mind-set of the person fishing it. Ponds don't generate the misgivings or the infinite reach that lakes or oceans or even the tumble of rivers arouse, but, complicit in the fishing and hunting exploits of man, they are forever an invitation to adventure. Ponds act as canvases for the sun, landing strips for the moon, and here in the South, a sanctuary for snakes and long-necked birds, softshell turtles, and alligators.

Every child has a pond he visits, if not physically, certainly in his dreams; a pond that, if he is lucky, he returns to as an old man. A few years ago I asked that my ashes be spread on the pond alongside my wife's (if she so chooses) or the ashes of my dogs (if she doesn't). In a wet year the pond's overflow eventually concludes in the Gulf of Mexico, and from there to countless places I have never been; an unspoken promise of adventures in the thereafter.

Pond making was once a sacred Eastern tradition. Monks carved ponds beside their temples and shrines in an attempt to mirror the universe. I choose to spend the final years of my life next to one pond among a thousand other ponds in northern Florida, a simple body of water without aspirations other than to reflect the stars and remind me of all the fresh- and saltwater destinations that have shaped my life.

SPRING POND

In April, after weeks of uncertainty, cold nights give way to temperate days. Blackbirds are sprinkled all over the trees next to the cabin. The live oaks, in the process of shedding last year's leaves, are crowned with fall colors in spring. For a few weeks they lose their stately allure as the dying foliage gradually gives up its hold and drifts to the ground. The inside of my boat is covered with dried-up oak leaves and the male cones of spent loblolly pines. Imported fragments of sunlight excite spiders into feeding inside webs filled with mayflies.

Overnight a family of otters mug the pond. They take turns pushing their heads and shoulders out of the water and staring at me as I sit on the deck watching them. Their black eyes don't bear the prejudice some men hold against them for eating "their" fish.

The otters are tentative at first. But a few hours later, having established that I am harmless, they return to the merriment of being otters. They chase, nip, and tenderly embrace each other, roll on their backs, tear into the bass and grass carp, and take naps under the dock, living as I wish I lived.

Since I can replace fish but not otters—and because I love watching them watch me—they are off-limits.

"But, they eat all the big fish," says Bill Poppell, my friend and farm manager.

"Don't care," I reply. Not the answer he had hoped for.

The grass carp—the largest and slowest fish in the pond, idiotic-looking vegetarians who are eaten, revered, and cloned in long pen strokes by Japanese artists—are going to take, as we say down South, an "ass whupping." After the otters depart for other ponds, I find partially eaten carcasses for weeks, the meat and skeletons abandoned at the pleasure of the otters for turtles to launder. The corpses all bear teeth marks at the base of the head.

Because they cut down the cypress trees and undermine the dam, the beavers that show up later in the year do not enjoy the same consideration. They are trapped. I don't know where Bill releases them, but I am told by some of the locals that beaver tail is tasty.

A steady wind yaws the bow of my jon boat off its intended path. Clouds of purple martins, resembling dandelion seeds, swirl in the breeze and leave dimpled reminders of their bills on the water. Yellow pine pollen wrinkles the pond's eyelids. A pair of small blue herons, bearing the stoop of old men, hunt together side by side in the reeds. The wood ducks squeal like socialites.

I maneuver the boat around logs that spent the winter underwater but have emerged with the change of seasons. I believe in the mystery of logs that rise and fall at the inducement of the moon. I believe in the purity of piano concertos, in the outrage of herons. I believe in those rites of nature that propose mystery and tears. I believe what my friend the painter Russell Chatham says: "There is no art without tears."

Once, a long time ago, below the coastal cliffs of northern Scotland, inland from the Isle of May, I drove past a pair of puffins, stocky birds wearing black-and-white feathers, short wings, and large, parrot-like orange-colored beaks. One of the birds had been hit by the car ahead of us, and its mate was running frantic circles around the inert body, pushing and probing it, wishing it with its wondrous bill back to life. Then it stopped and stood forlorn, its beak resting on the breast feathers of its mate. Tightly framed through the vehicle's window, the scene was a still life of agony. Twenty years later I imagine the puffin's tears and relive the incident as if it happened yesterday.

In May I watched the remaining two young geese of an original clutch of four learn to fly. They wobbled low and tippy over the pond at first and then flew with more courage until they were ready to leave. On the morning of what might have been their last night in northern Florida, I found the two goslings dead on the shore, the mother's body dragged fifty yards to cover, her breast devoured. Ever since, I have wondered what killed the birds. Geese sleep on land. They are birds of size and will fight. It must have been an animal of strength, sense, and speed. It would have killed the mother first before turning to the young birds, who no doubt reacted with terror, hopping about, wings spread, honking and behaving like carnival queens. An alligator would have pulled the mother goose into the pond. A black bear would not have bothered with the goslings. Maybe it was a pack of dogs, but I saw no evidence of that. I would like to believe that a female Florida panther had herself a party down by the pond that night, but I know the odds of that being the case are slim to none.

In the same spirit that I waited for the geese to leave the pond, I often watch baby wood ducks swim single file behind their mothers in late spring, their tiny spinning feet and sugar-spun bodies straining with the effort of keeping up. Predictably, first I see a swirl of water, then a pockmark where a duckling had been swimming an instant before. Bullfrogs and moccasins eat hatchlings; snapping turtles, alligators, and large bass take baby ducks. By the time the young birds are ready to fly, the clutch is down to three, maybe four.

A while back I saw a lone small wood duck, abandoned by its mother. The chick swam randomly in circles on the surface, acting out the motions of its windup, bright yellow plastic counterpart in a bathtub. I knew it would not live through the night and that its death would be sudden and certain.

Observing myself through the eyes of another species begs a question regarding overvalued consciousness. I am not the work of a god but of the inexorable climb of evolution through thousands of millions of years from stardust, the carbon from exploded stars.

Half an hour after sunset, I sit alone at the window of the pond house thinking about the salmon-colored light borrowed from the clouds and spread by the breeze over the water. Inside the curl of ripples raised by the wind, the light is fractured. While I wait for darkness, silver water sparkles between the reflections of the pine trees rising from the bank. The wind drops and the pond releases its hold on the waning radiance. I watch the sky lose its position.

From my world here in the cabin, I gaze through the window at the pond's inscrutable surface. Out there, looking in, it is hot and mysterious and alive. Frogs on the glass, splayed tight to the window, hunt moths attracted to the glow inside the cabin. This evening the frogs are quiet, satiated perhaps.

One last time before total darkness, the breeze rises again and shapes dozens of disturbances over the surface that look like rising fish. Sounds that herald the night emerge from the shoreline and stimulate the imagination.

TO BUILD A POND

Planet Earth brightens the screen of the computer on my desk in the form of a blue orb revolving in space. After choosing *The Pond* from a preset menu on the opening page of Google Earth, I sit back and watch the effects of an omnipotent eye plunging ever faster into northern Florida. Unstable from the speed of its fall, the lens blurs before alighting on what I know to be a small body of water situated at latitude 30° north and longitude 84° west. Instantly the picture tightens into focus, a revelation that is not so much a product of silicon as it is magic.

The shape of the pond, in reality an impoundment, resembles the shadow of a woodpecker, a stylized cartoon of a bird with a beak. The land beneath the water had once been a bottom, packed with magnolia and gum trees, water oaks, hickories, and climbing vines all reaching skyward, competing for sunlight. In 1974 Fanny Malone, a lady from Tallahassee who owned the farm I now live on, hired a crew to clear-cut ten acres of hardwoods at the bottom of the hill her house had been built on. She and her second husband, William Malone, had a dam built at the drainage end, a clay impediment to keep the water from escaping. Fanny had married young into a Scottish family whose ancestors had migrated south from Scotland to the town of Quincy, located between the arms of the Ochlockonee and

Apalachicola Rivers at about the time Andrew Jackson began driving the Creek Indians off their lands.

In 1990, sixteen years after Fanny supervised the construction of the original dam, she sold me her farm. A year later I raised the weir by twelve feet and widened its base by twenty-nine feet, thus encouraging water to back up into a pair of clay ravines and a secondary hardwood bottom I had razed at the swamp end. Water stole into seventeen more acres of land and altered the water's edge so that from space the pond now looks like a bird.

I have read that the size of a pond should not exceed ten acres. After that it should be referred to as a *lake*. By raising the dam, I elevated the pond's social status, but because I like the resonance of the word *pond* better than I do *lake*, I refer to and think of the gathering of water at the bottom of the hill as *The Pond*.

The expansion of the original body of water into the clear-cut hardwood bottom was the result of a set of measured choices dictated by a sense of how far I wanted to throw the water. I started the process by using a transit, an instrument familiar to builders but foreign to me. According to Charlie, the equipment operator/contractor from Georgia, and Bill Poppell, the flags we planted along the path of the hypothetical pond would mark its perimeter, to the inch. The exact amount of dirt we would have to add to the existing dam in order to satisfy the evolution of the pond was determined after two days of working back from a point of desire to the existing structure.

I held the measuring stick where I was instructed, ratcheted my thumb up and down the numbered scales, and moved it a foot here, a foot there, back and forth all around the edge of the wreckage left by the timber crew who had clear-cut the swamp. When we were finished, I looked back at the path of orange flags that we had planted and felt only skepticism. The simplicity of the method, similar to other senseless science class experiments of my youth, had left me unconvinced. Like a doubting Thomas, I would have to see it with my own eyes.

The next step was siphoning ten feet of water off the base of the original dam in order to dig up the original, metal overflow pipe. This was accomplished by a 5-horsepower, two-stroke generator pump that ran from dawn to dusk, emitting loud ratchet sounds that were soon imbedded into the daily routine of the farm. While their new quarters were being expanded, the existing population of largemouth bass, catfish, perch, and bream were gradually concentrated inside five acres of water.

During the three weeks of drawdown, fresh clay on the south side of the original pond was gradually exposed. The sound of the pump was replaced by the sound of a 350-horsepower bulldozer and a track hoe named Gertie. A dirt pan attached to the bulldozer peeled the hillsides of dirt, two inches at a time, until enough had been collected to be delivered to the dam. The good dirt, thick and heavy and clay colored, was spread, worked, and packed tight to mold the edifice. For months the dozer and the pan shuffled across the structure in a lazy figure-eight dance, cutting, scooping, and dumping.

Building a dam with a 350-horsepower dozer, a track hoe, and a rented pan is an unthinkably slow process. Brutish perseverance is the only approach. A few months into the project, Gertie took to displaying her age. Choleric, she sighed and smoked and shook and blew hoses, dumping fuel into the water. Added to her misfortunes, each time it rained her canvas turned into a bowl of clay gumbo she was unable to work in.

And then one day there was no noise. The pond was silent, and it was over. Seven months after the first bucket of dirt had been scooped up and then emptied out on top of the old dam, the structure was pronounced finished. The new dam stood terra-cotta proud, large and naked against a backdrop of fall foliage.

Charlie beamed and said, "We sure enough toted some dirt."

Rolls of muslin cloth once used to shade tobacco leaves from the sun were salvaged from an abandoned drying barn and used to clothe the structure. The close-fitting weave would allow grass to grow between its stitches while keeping the clay from running down the slope during a rainstorm.

After months of asking inane questions, skulking, and writing checks, I was told that my final duty consisted of exercising patience—great patience—because the pond was going to take some time to fill. It turned out that the mind-numbing process of watching rainwater fill a twenty-seven-acre hole would demand Biblical commitment and the resolve of an old dog fucking.

In the beginning, fingerlings of fresh water trickled out of the clay, coursed down the sides of the dry pond, and puddled at its bottom. As the puddles grew and the rains fell, the springs eventually lost their collective eagerness. Day after day, month after month, as water slowly seeped into the clear-cut swamp that was the addition to the original pond, I walked its edges and observed what looked like an abandoned construction site in Biloxi, Mississippi. In their distinctive manner, the timber crew had ripped out the moneyed trees and left behind the usual backlog of empty cans, shit wrappers, and grinding wheel marks. The outcome of the clear-cut was a collection of ignored logs, immature trees, the crowns of harvested hardwoods, jagged stumps, and bad feelings, all of which I harbored.

The unsightly trash and untidy landscape stood for months as a daily reminder that what I had done (turn a dank, sweaty, living bog into a single body of fresh water) had probably not been ecologically sound. Without a permit from the Water Management Office, which I never would have been awarded, it was also illegal. By definition, a dam is a scientific thought run afoul.

The lowlands that I clear-cut to expand the size of the pond had once been part of a dark, narrow slough crowded with deciduous trees that coiled snake-like for miles through the woodlands of long-leaf pines that cloaked Gadsden County. Historically the swamp had been a setting from which to ambush bears, deer, and turkeys feeding on acorns. Woodlands that—decades after the last Creek Indian had been relocated to Oklahoma and the last Seminole had made his escape to the Everglades—fell to the corporate greed of the timber barons who, in sixty years, the average life of a man, cut two million acres of standing long-leaf pines in the Southeast, the pride of Southern forestry.

Waiting for rain is what people do on farms. Delays in delivery are punctuated by back-slapping optimism, sometimes prayers (particularly in Baptist congregations), and eventually sighs of resignation. Eight months after the last scoop of dirt had been applied to the dam, a tropical storm inbound from the Gulf delivered eleven inches of rain on the county in three days and completed what had been until then a process that had found me staring like an ox for almost a year at water creeping toward a small army of fading orange flags.

The rains came and stayed. On the first day the soil soaked up all the water. On the second and third days, great sheets of water poured from the hillsides into the pond. Late that afternoon the mist rose from the clearing and took the light with it. On the morning of the fourth day, the rain stopped with an abruptness common to our latitudes.

When the sun emerged, I woke to a sonata of light on water at the bottom of a green hill. The pond was full, and the sudden

sight of water where there had been only weeds growing out of
a black swamp was foreign. I looked away and looked back half
a dozen times. It was as though a vineyard had sprouted wine
bottles overnight. I stared at water that would shelter frogs that
afternoon and mirror the stars by nightfall.

The water lapped against the wire stems of every sun-
bleached orange flag, proffering up an exhibition of engineering
perfection and, to an ill-informed romantic, a further instance
of magic. The pine trees closest to the pond stood tall next to
a new, unfamiliar habitat. Their long reflections in the water
revealed the elegance of their form.

As the storm subsided, seasonal shorebirds, godwits, sand-
pipers, dowitchers, and every species of heron took to the
water's edge. Within a few hours there were birds everywhere
running the brand-new shoreline, spearing invertebrates, and
asserting authority over a newly born landscape.

The storm water overflow whirled through a thirty-inch
riser cemented to the bottom of the dam and poured through
lowland hardwoods into my neighbor's pond. From there the
water found passage under the bridge that crosses the blacktop
of rural Route 159 in Gadsden County. Soon afterward, ele-
ments of my pond fell into the Ochlockonee River.

The flow of water courses three miles from my place to the
river and another sixty-five miles to the Gulf of Mexico. From
there, and in time, fragments of the pond drift, urged by the
harmony of currents and wind, to all the seas and oceans of the
world.

My pond is located west of Havana, Florida, baptized after
the first city of Cuba, and east of the village of Midway, so named
because it rests halfway between the towns of Quincy and Tallahassee, the state capital. Slightly more interesting is the fact
that the pond is located about halfway between Sopchoppy in
Wakulla County to the south, and Whigham, Georgia, to the

north—quiet Southern towns except during one week in January and April, when they are transformed into monuments of low-rent distraction.

North of the pond, Whigham (population 584) is located in Grady County, an agricultural region disposed to growing peanuts. The town was built in 1905 and to this day hosts a rattlesnake roundup on the last Saturday in January. In the days leading up to Whigham's featured weekend, men dressed in camouflage clothing, shadowed by their similarly clad progeny, head for the woods individually or in teams to smoke out or fume out (by pouring gasoline down a den) rattlesnakes. Hundreds of them, netted as they abandon their burrows, are transported to an outside arena the size of two double-wide trailers. Families stand three deep at the wire-fence partitions to experience the thrill of seeing up close hundreds of live reptiles ranging from two to six feet long. For the most part the rattlesnakes rest quietly on the ground, but their numbers and immediate proximity are enough to rekindle the childhood nightmare of falling into a pit filled with writhing vipers. Some of the children whimper. Their mothers pull them close.

The Whigham Rattlesnake Roundup merchants set up temporary booths that feature rattlesnake-skin hatbands, bags, belts, jockstraps, boots, and condoms. There are rattlesnake skins, fangs, and head mounts (mouth open: striking posture). The culinary mainstay that day is fried rattlesnake and funnel cake. Cowboy boots prevail, wife-beater shirts are code.

South of the pond lies the town of Sopchoppy (population 498), founded in 1894 and built eight miles north of the Gulf of Mexico. It is known in these parts of Florida for its annual Worm Gruntin' Festival, an event as important to some as the last square of toilet paper is to others and to which thousands of folks from Tallahassee and southern Georgia attend every year on the second Saturday in April. The festival weekend

attracts Southern families interested in bettering their grunting skills and expanding their social interactions. It is also a
welcome excuse to eat cheap deep-fried food. Picture a sea of
faded madras shorts straining against pale thighs and Sharpie-
detailed varicose veins. White T-shirts cradle low-hanger tum-
mies. Baseball caps and shrieking children add to the gallery.
The festival features a 5K race advertised by the admonition,
"Come run with the worms!"

Worm grunting, a common practice in the Apalachicola Forest before worm farms commercialized the process, is
the art of luring earthworms out of the ground by rubbing a
piece of flatiron over the head of a cherrywood stake (known
as a stab) driven into the ground. When executed with know-
how, the rhythmic back-and-forth motion of metal on wood
produces a noise comparable to that made by a blind eastern
mole—a gourmand of invertebrates—on the prowl. Scores of
worms within earshot of the imitation mole call predictably
rise in a hurried manner out of the darkness of their burrows
to the safety of daylight. There, and insensitive to their plight,
the grunter collects the worms and drops them into a tin can.

The art of worm grunting is practiced by tourists but has
been perfected by a handful of aging grunters, some of whom
have been rubbing metal on wood for decades. The fruit of their
specialized labor, a coffee can of lively earthworms, sells for
twenty-five dollars.

"You are born a worm grunter," a desiccated old man said
to me once, sipping on his beer in Bullwinkle's Saloon. "It's a
hard-ass thing to teach."

The pond is now a small, sedentary body of water that travels only when impelled by tropical storms or, conversely, by
droughts. It is water that does not personify the intrinsic property of motion: the quest of rivers, of tides, of clouds. The body

of my pond is young, its history undeveloped, its sensibilities insensitive, and its wisdom embryonic. But it is water, and since the composition of my blood is saturated with water, my kinship is immediate and profound. Symbiotic by nature, the pond is a pool of silence, beauty, and diversity to me every day. More down to earth and akin to the flower girls I used to date in Key West, her terroir is memorable. Smooth or fractured, fresh or salted, the presence of water reassures me; the medium is familiar, finite, a stage pleasing to play on.

That first morning after the storm had moved north and made a statement out of a hole, and every morning since, this small body of water is where I come to get away from human interference and observe, on a small, contained scale, nature's simple and effective protocol. My faith in the ecosystem affirms the reverie and the mystery that water evokes, and the toughness by which nature subsists. Water unpolluted by man is a joy to the eye and a sanctuary from which to observe the genesis of life, not to mention the inconsequentiality of one's own. Inside the pond there is no vanity, no malice; there is no greed. Fish don't have social aspirations; they have no egos.

LA FRANCE

Sixty years ago, when France was still reeling from the horrors of the Second World War, and when Allied bomb craters littered the forests of Normandy, I lived in a seventeenth-century castle that had been constructed on a man-made island situated between two rivers in the southeastern region of Normandy. Built of pink brick and mortar, the castle was on sunny days reflected in the water of the moat that surrounded it. Originally designed to beautify the appearance of the castle, the moat had also provided nominal protection to its inhabitants against the wolves and bandits that roamed the French countryside.

Water released from the bordering rivers irrigated the rocky soil of the land I was raised on, and over time it had chiseled a warren of brooks and streams that flowed into the moat and the water gardens, the ponds and basins, and fed the weirs that water engineers had designed hundreds of years earlier.

The moat that surrounded the castle extended outward to include to the west a large gravel courtyard where horses and later cars deposited their charges, and to the east a square of mowed grass, shaped yew trees, and flower beds that stretched farther than the flight of an arrow. North and south of the castle, two rectangular basins, permanently shaded by the lean of centuries-old oak trees, had been carved into the black earth of Normandy.

When I was ten years old, my father stocked the southern basin with trout for the enjoyment of his guests. He referred to the fish-filled basin as *"Le basin des couillons"* (*couillon* loosely translates into "numbnuts"). Anything dropped into the water, including a bare hook, was instantly set upon by the pellet-fed, soft-to-the-tongue rainbow trout imported from *les Ameriques*, trout known for their hardy nature and undiscerning appetite. A typical guest, much like a character in a Moliere play, would, rod in hand, dance in wonder on the banks of the basin and after each catch invariably thrust his chest forward and with marvel spilling from his lips say things such as, *"C'est formidable!"*

My father had fought and been wounded in both World Wars. In the First he was shot off a horse, lance in hand, during an ill-conceived charge against a cavalry of Germans who cleverly dismounted and fired at the storming French Dragoons. The bullet grazed my father's temple but caused no damage other than a lifelong dent in his forehead. He was wounded a second time twenty years later by a German pilot during a single-engine air combat over the town of Évreux, fifty miles west of Paris. This time he spent eight months in hospital before escaping and making his way to Great Britain via Gibraltar in a repossessed German submarine hunter. Understandably, my father did not eat sauerkraut or listen to Richard Wagner.

When the politicians finally called a halt to the killings and carved up the lands they deemed valuable, my father, like the many thousands of other men who had fought for a cause and withstood the atrocities, was confronted with the numbing reality of normalcy. An engineer and keen storyteller, he prized the ridiculousness of men who took themselves seriously, and defined ego as a reflex of ignorance and delusion.

In those days, with the notable exception of Charles Ritz, few Frenchmen were accomplished fly fishermen. Casting was engaged entirely from the wrist, with uninspired twelve-to-two repetitive energy. The results lacked style and substance: Flights of feathers and hair twirled around the shanks of small hooks rose and dipped ungracefully in a medium of air and hope, and invariably landed far from the mark. However, since the trout my father had stocked in the basin would rush at any and all disturbances, unless they hooked themselves or each other, his friends always caught fish. At the dinner table, in recognition of the physical demands, not to mention the mental stress imposed by the sport, my father would raise a glass of wine and toast his friends' accomplishments.

I didn't care about the sport of it. A fish was a fish, and I liked all fish, the bigger the better. The pale renditions of the rainbow trout that did not die of infectious maladies nor were caught by my father's friends were large fish, and I took to tickling them, an art taught to me by an older man with long arms who drew pleasure in delighting fish and women, in that order. He pointed to where I should lie on the grass next to the sluice gates and demonstrated the correct angle my hand should be inserted into the water to feel for quiescent fish. Even though the blood of their wild antecedents had been bred out of their genes, the basin trout nonetheless sought the whisper of moving water. Somnambular in its contentment, a napping trout would allow me to run my fingers along its flank, past the feathering of its fins, and entertain it into a condition of misguided abandon. When my fingers reached the gill plates, I would close them and yank the fish out of the water and onto the bank.

Tickling stocked rainbow trout in the basin was but a prelude to my later attempts to hand catch the warier brown trout that resided in the rivers and streams that threaded through the property. I spent hours of my youth lying on grassy banks trying to avoid the stinging nettles, my head cocked to one side, my eyes closed, feeling for a slick, cloudlike, almost imaginary body lingering on the edge of the current. In contrast to the domestic fish that swam above the mire that coated the bottom of the shallow basin and ate the residual chicken pellets dispersed daily by the gamekeeper—which blanched their character and flesh—the odd wild trout I managed to tickle in the cold, clear water of the river was rushed to the kitchen and cooked *au bleu*. The recipe demands that the cook insert the trout, fins aflutter, moments after removing its guts, into a court bouillon. Traces of white vinegar in the broth turn the trout's skin a lovely transparent shade of blue. The wild river fish was served alongside a lemon-butter sauce, its pink flesh evidence of its taste for freshwater shrimp, its firmness a tribute to a species that has survived without assistance for millennium.

The moat and its channeled extensions were six feet deep and filled with fish. On the castle side, a brick-and-cement wall protected the building; sixty feet away, on the open side, grass banks rolled up out of the water to meet graveled pathways. The small fish exploited the shade of the walls and bridges for protection; the big fish made use of the shade as a stage from which to initiate their offense. The eels that shyly retired by day into the soft, dark mud at the bottom of the moat were caught by a worm-baited trotline set at nightfall. The imperative to this method of fishing was that the line be retrieved in the dark. At sunrise the eels would start fighting the hooks until they ripped them out of their stomachs.

The innocuous roach fish that lived in schools and waved orange-colored fins were caught on cane rods outfitted with bobbers and tiny golden hooks baited with miniature bread balls. Small roach were occasionally fried, but since each one had to be cleaned and scaled, it was rarely worth the effort. In any case, they tasted like every other little fish in the world cooked in hot oil.

The perch, a percentage of which reached a pound in weight, were fighters. They chased speckled spinners or took small, energetic red worms. Once hooked, they showcased their strength by pulling line off the spool of the reel. Fillets of perch were sautéed in butter and sprinkled with fresh lemon juice and minced parsley. If caught in a current of water—water that flowed in or out the nearby streams and rivers—the meat was sweet and tasty. Caught in the dormant waters of the moat or its adjoining basins advanced the flavor of inactivity imbedded in its tissues.

Small schools of chub, black tailed and vegetarian, often swam just beneath the surface. They could be enticed to strike an imitation black housefly by landing it on their nose. Chub were oily and not good eating. Thought by French river keepers to be damaging to the health of other more palatable fish— a false notion based on the hubristic concept of managing nature by purging the competing or subordinate species—all chub pulled out of the water were left flopping on the bank, free lunch to crows and magpies, who plucked out their eyes to access their brains.

Pike were the main piscatory and culinary attraction. The largest ever caught in the moat weighed twenty-three pounds and survived in a yellowing picture that I discovered in my father's library. The date reads May 18, 1896, and the photo features a grinning, bearded, middle-aged man wearing a cap and suspenders buttoned to high pants. He is holding the beast at waist height with his left hand inside its gills, and gripping his bamboo rod with his right, the butt end planted into the ground next to his feet and the tip shoved out a stiff arm's length away from his body. The gills of the lifeless pike flared like the pink hats of ladies on derby day.

The methods of catching pike varied, and some were more successful than others. On summer days those fish that rose to the surface of the moat to bask presented tempting targets to one inclined to shoot a .22 long rifle. Shooting pike was a practice I indulged in for a few weeks after I received a gift of a real gun from my father, when, for a moment in my life, I shot everything, including fish.

After that moment passed, I learned to cast big silver and gold spinners into the shadows thrown by the walls of the moat, and gaudy wooden plugs brought back from America. Once I learned to cast a fly, my favorite was a white deer hair and Mylar streamer. But no matter how much effort I exerted, neither the plugs nor the flies worked with the consistency necessary to build confidence.

Before a storm, when the clouds bowed low, or sometimes when it rained and the light was gray, something big and determined would push across the water and take the lure. With the taking, the fish would change direction and try to pull the rod out of my hands. If I did all the right things, I would win the fight. But more often than not there would be no fight, because the shadow behind my lure would come and go without me seeing it, or, worse, I would miss the strike and instead be left staring at an insulting swirl of water.

Pike are by nature moody creatures with difficult dispositions. Often I would see them as a shape of something that didn't belong, or catch a glimpse of an outline, a shadow, the beat of a tail. Other times they would sun, inches under the surface, and I could count every scale on the fish's back.

I learned to work on stratagems, setups (which often involved crawling on my belly to the side of a bridge or behind an oak tree in order to stay hidden), casting angles, and retrieves. Then one summer I swam a live bait a foot in front of a pike—a technique learned while saltwater fishing in the Bahamas—employing a small perch in such a way that it came into the sight of the pike seemingly wounded. The free meal proved irresistible. The largest fish I caught in the moat using live bait weighed fourteen pounds, but I caught many six- and seven-pound pike, whose size delighted the cook.

A big man from Burgundy, the cook liked to present them whole to the table, poached in a court bouillon made of white wine and water, a dash of red wine, a carrot, an onion, some shallots, garlic, salt, peppercorns, and a bouquet garni: a bunch of parsley and thyme and a bay leaf tied together. First he brought the broth to a boil and simmered the poaching liquid for thirty minutes. Then he added the pike. When the fish was tender, he removed it, drained it, and wiped the skin off its body, exposing the delicate off-white flesh. The cooking liquid was reduced and clarified and then encouraged to congeal with the addition of gelatin. Before the liquid had time to cool, the cook stamped thin slices of lemons or truffles up and down the back of the pike for decoration and poured the broth over the fish to mask it. When the broth chilled, it formed a thin gelatinous film that highlighted the trimmings. He served the pike on a platter alongside a cold remoulade: a simple homemade mayonnaise to which he added capers, minced shallots, chopped pickled gherkins, and mustard.

Because the cook and my mother and father voiced their approval of my contributions to the table, I came to believe by the age of twelve that my fishing misadventures were sanctioned. This led to further adventures into a world that increasingly and mercifully spared me the interest of my parents, or so I thought. The high points of these activities occurred at night when I would slip out of my bedroom to meet Michel, a friend from the village. We would rendezvous far enough from the castle so that the light from the beam of our flashlights was masked by the ancient trees that cast colossal night shadows on our fishing grounds. Full moons dropped puzzling light on the floor of the forest, but we preferred new moons, when our flashlights were a traveling necessity.

Michel brought our weapon—a narrow, four-pronged gig—and a gunnysack that we used to house our quarry. We hunted eels. We hunted them in ponds and on the edges of eddies and miniature oxbows where the mud had settled. Taking turns, one of us would aim the flashlight a few feet from the water's edge while the other carried the six-foot-long spear at the ready. Our sneakers gurgled water as we looked under the surface for the telltale motion of what we knew to be ferocious brown bodies led by spade-shaped heads, wide grinning mouths, and tiny black eyes. The eels cruised the edges of the shorelines and hunted for worms and frogs, baby ducks and crayfish, smaller eels, tadpoles, and fish, all the while keeping their menacing, undulating bodies between their prey and the safety of deep water. Eels are fast strikers; tenacious, short, vicious apparitions that even when dead and limp terrified my sister, her friends, and many of my father's guests.

It was preferable to strike an eel while holding on to the gig with both hands, but if the target was out of arm range, we would throw the harpoon like a spear and follow it into knee-deep mud. The eels, often three feet long and as big around as our arms, were solid chunks of meat and muscle. When harpooned they would contract and coil, open mouthed, around the gig's wooden handle. Neither of us was ever bitten, but each time we went in the water after them, we thought about the eel and its small teeth closing on our flesh with the same conviction it displayed when, impervious to pain, it ripped an offending hook out of its stomach in an attempt to detach itself from a trotline.

On a good night we might gig three yellow-bellied eels. One was the norm.

Michel and I knew that mature eels left our rivers each fall for the Sargasso Sea, south of Bermuda, a three-thousand-mile swim across the Atlantic Ocean. Once they reached their destination, they mated and then died. We looked every year for what had been described to us by old-timers in the village as legions of bodies slithering across the fields and meadows of France en route to those rivers that would carry them to the sea, but we were never privy to such a sight.

The delicate and delicious flesh of the eels has inspired men to study them. Aristotle incorrectly declared that like worms they "Grew out of the guts of wet soil." In fact eels are a catadromous species, fish that mature in fresh water but spawn and die at sea. The larvae conceived after their arduous journey know to follow the currents of the Gulf Stream and return to the rivers of Western Europe. A similar but shorter journey applies to the eels that return to the rivers on the East Coast of the United States. By the time the European eels enter fresh water, a one- to three-year swim from their place of birth, the larvae have developed into glass eels, so called because of the transparency of their bodies.

Glass eels are netted by the millions in the rivers of Europe and sold to the Asians for astronomical prices—up to five thousand dollars a kilo in Hong Kong. It has been said with debatable verisimilitude that Japanese men insert baby eels into their urethra to combat impotence. This demand on eels, like our other gluttonous claims on wildlife, has depleted the stock worldwide. In the United States, glass eels are usually smoked; in Great Britain they are jellied and potted; and in Spain they are fried in olive oil at five hundred dollars a plate. Parasites and diseases caused by uncontrolled aquaculture have added to the demise of wild eels and greatly diminished the numbers of strange and sinister creatures my friend and I would stare at in wonder as they emerged from the mud on those dark summer nights in Normandy half a century ago.

My parents loved winter eel stew, an enduring example of Burgundy comfort food. The cook would hang the eels alive from a wooden beam outside the kitchen and, using a funnel, pour white vinegar down their throats, a purging operation he insisted cleaned them of an otherwise lingering odor of mud. Once the eels were dead, he would make a circular incision at the neck of each one and peel its skin off in one fluid, downward motion. He cut the pink meat into sections and marinated it in red wine and herbs for a few hours before dusting it with flour and sautéing the roundels in butter. It was then simmered in the marinade alongside carrots, small potatoes, and previously blanched salt pork.

The cook, an opinionated man under the best of circumstances, rejected eels that had been pierced by prongs, even if they were alive when I handed them to him. "*Je ne veux pas voir de trous dans mes anguilles*," he would say loudly, "I do not want to see holes in my eels."

I learned how satisfying it was to reply, "Oh, go fuck yourself," behind his back and under my breath.

On spring nights we hunted for brown trout in the streams and rivulets that fed the fields and pastures of the property. Edgy from being in the crosshairs of pike, herons, and otters, the trout were preternaturally aware of our slightest misstep. A stumble, the snap of a twig underfoot, or the briefest shadow on the water would send them scurrying from their relative shelter under logjams or fallen trees, the calm water behind boulders, or the occasional hunk of metal (mementos of the war) to the safety of deep, fast, running water.

Our nocturnal pursuits on the edges of the streams were demanding and precise. We walked lightly on grazed pastures and embankments in the dark, followed at a distance by cows curious about our intentions. When one of us would enter the water, the cows would brutishly bugle their displeasure. Illuminated under the yellow beam of our flashlight, the signature brass colors and the halos that softened the black-and-red orioles of a brown trout's flank were difficult to see. It took some time to decipher a flurry of fins, or the flinch of a tail against gravel, motion inside motion, the opening and closing of a mouth, a gill, as the targets we were after.

Gigging trout was illegal and cause for serious fines in France, and the river keepers were allowed to patrol on private property. Although we were never apprehended or even chased, it was out of the question to bring home a lanced trout to the chef. Our nightly victims found their way into a gunnysack and were transported to the village by Michel, whose mother, a good country cook, loved baking trout in cream and Calvados.

My father knew what I was up to on those long summer nights, but he feigned ignorance. One night he must have followed us, for when I sloshed my way through my bedroom door at two o'clock in the morning, he was sitting on my bed.

"Have a productive evening?" my father asked me. Having scared me half witless, he laughed. "Just be careful," he said. "Don't walk that gig into Michel's ass. Or worse, in your foot."

By far the most productive and exciting poaching adventures were when we electrocuted the moat, an activity relegated to those times that my parents were out of town. Our weapon consisted of two metal plates soldered to a length of coated wire at one end and an electric plug at the other. Michel and I split the wire and tied the plates to the bow and stern ends of a twelve-foot wooden boat. Then we moved from one section of the moat to the next, threading the wire through the windows and inserting the plug into the wall sockets of one room after another.

We took turns rowing the boat and used a wood-handled net to gather fish that invariably boiled up between the poles. The chub, perch, and roach caught between the terminals froze mid-water, but the eels ripped out of the mud and kept fighting the ignominy of their predicament until one of us netted them. We never caught a pike.

The castle was wired with 220V current, which posed major concerns, especially when the boat started to fill up with moribund fish, some still moving, particularly the eels, slithering over the bodies of the rough fish, rigid in death.

One day one of the sockets overheated, and flames licked and blackened the castle walls. This time matters did not go as smoothly with my father. I was sent to my room with a sore butt, a pad of paper, and the assignment of writing apology letters to my mother, my father, my sister, and every grown-up in the house, including the maids, the butler, and the cook, all of whom may well have died had the fire developed.

Fishing to music started during those night hunts, beginning with the heart-stopping fervor of Peter Wolf and continuing a few years later to the songs of the Beatles, the swagger of the Stones, and the lamentations of Gypsy singer Mouloudji. When I was thirty and living in the Keys, I fished to the melodies of my friend Jimmy Buffett, then to Santana, Clapton, and even the Bee Gees.

Now that half a dozen decades have come and gone, I immerse myself in the mostly silent world of nature. I don't wear earphones or read books when I fish, because the wind, the waves, the wakes of hunting fish, the circular impression of a bream rise, and the consequence of bird shadows on water all carry a melody. And just as some hear the music of the choir, I listen to the beating heart of souls whose struggles, loves, and deaths are indistinguishable from our own.

THE SOUND OF FISH SWIMMING

On the bank next to the cabin, a twelve-foot aluminum boat is pulled up onto two parallel beams of treated pine lumber cross tied with steel rods and PVC rollers. A thirty-pound-thrust electric engine hangs on the square bow, two swivel chairs sit front and aft, and a sheet of outdoor carpeting conceals the floor. Speed launching is a matter of disengaging the sprocket from the cogwheel and giving the bow a shove. Three minutes after making up my mind, I am fishing.

A long time ago a dog trainer friend of mine told me that he always took his dogs fishing to teach them patience and introduce them to a world that had nothing to do with birds, whistles, and guns. He has moved on now, but I still follow his advice and take my dogs with me in the boat, particularly when they are young and impulsive. Being surrounded by water without being allowed to jump in teaches them trust and patience. Over the years they, in turn, have trained me. These days I have a hard time refusing the willing looks and tail wagging that occurs every time I grab a rod and push a handful of flies or worms in the pocket of my T-shirt. Admittedly, I love the company of dogs.

For a while I took all three dogs—two French Brittanys and one English cocker spaniel—but although it was amusing when the fishing was slow, the trip invariably turned into a clusterfuck when the bite was on. My twelve-foot boat isn't built for the jostling of canines, or worse, their wishes to be the first to gnaw on the tail of a bass dangling from the fishing pole. To their credit the dogs portray pictures of canine contentment: mouths open, tails in motion, paws resting on the narrow gunnels. A pastoral setting until the fish start biting, or when Louis, the young male dog, chooses the bow of the boat, three feet from where I sit, as his outhouse.

Keener than I, the dogs focus for longer stretches of time on the rubber-legged action of the small black-and-yellow popping bugs swimming across the water's surface. Often I have seen Heather, the little white-and-tan cocker, tilt her head to one side just before a fish strikes the lure. The bitch can hear the sound of fish swimming, and for that reason I take her fishing more often than I do the Brittanys. She is smart, while the two bird dogs are simply enthusiastic. The older I grow, the more I appreciate smart.

It is clear to me, and yet a mystery, how dogs are able to sense a presence outside of what I would consider a normal range of perception. My house is more than a quarter of a mile from the county road, but all the dogs come to attention and start barking the moment a car or truck drives over the cattle guard at the entrance. Heather had refined this talent to sensing fish underwater.

Now that I've gone back to using a plug or spinning rod and plastic worms for bass and a light fly rod for bream, I mostly fish clad in shorts and a soft, wide-brimmed hat that thirty years ago I would have referred to as a geezer hat. Frankly I don't care anymore about what anyone else thinks about what I wear or what I do. My approach to sport is now one of utter simplicity, which explains my renewed fondness for cane poles and worms. I don't feel any kinship with the new breeds of anglers who fetishize the flex of the rod, the shape of the fly line, or the drag of the reel, and who, to combat the technocrat in their souls, go to the far extreme of what they refer to as consciousness—and to what I generously refer to as Dadaism—anglers who kiss the fish they catch as if the fish, God, or anyone else gave a shit.

When I threw the catfish food off the dock yesterday, the bream were tentative and sluggish, but last night the temperature remained above fifty degrees. This morning, in concert with the carpenter bees, the fish swarmed over the pellets and the bees swarmed the cedar planking of the pond house. The holes on the surface where the bream rose to the pellets and the holes in the wood the bees like to drill were perfectly cylindrical.

Spring is in bloom. It's time to go fishing.

With the sun behind a high layer of gray clouds, I gather Heather, turn my back on the noses of the two French dogs fogging the window pane, and push the boat into the water. With the help of the electric engine, we cruise in silence to the far end of the pond. The cypress trees are leafless and the banks leading down to the water are brown and thin of vegetation. The wind is light and out of the northeast, which will make for a clean drift back to the cabin.

I like to see nature flourish as it had for thousands of years before the Industrial Revolution. In most places nothing can be done to recapture the past, but here, since the small patches of cultivated ground and the clearings in the woods—common during the era of small tenant farms—do not exist anymore, I open the land to the sun, shoot a hundred doves and four dozen quail a year for the table, and feed ten thousand more birds, both local and migratory, that stopover at the farm. I apply similar practices to the pond.

Without harvesting a percentage of their residents, Southern ponds produce bass that multiply at an unsustainable rate for the size of their habitat and the food it produces. The water of my pond incubates and hatches millions of fish eggs a year. From the tiny percentage that survives, I cull five bass per acre, which keeps the waters healthy, the bass growing, and my tummy happy.

On the way to the dam, I thread the six-pound test monofilament that fills the spool of my spinning rod through a one-eighth-ounce conical weight that fits up against the eye of a 2/0 worm hook. I choose a hook with a built-in wire loop guard soldered to the shank. The worm will swim through the water without gathering the weeds, grasses, and plants that grow in its path. When the bass strikes, the guard will open, revealing bare steel.

In the spring, soon after the females return to deep water, the bass gorge in an effort to regain the weight they lost during the breeding season. I would normally fish a weightless worm and retrieve it through the cover next to the water's edge, waiting for a sudden elbow of water to form behind it, but, since the drought has expunged most of the aquatic plants from the shoreline, I have to fish deeper.

At the dam I thread the hook through the body of a seven-inch purple worm and push its plastic body up the shank to the eyelet. The end of the guard fits perfectly over the tip of the hook.

The breeze scatters pollen off the branches of the loblolly pines next to the pond, and a diaphanous yellow sheen settles on the surface inside the arms and small bays of the shoreline. The folks in town are sneezing and rubbing their eyes, grumbling just as they did when the weather turned cold in January. In August, when the temperature climbs into the high nineties and the humidity grabs them by the throat, they will grumble some more, and I'll go fishing.

Since I enjoy looking more than I do fishing, I prefer to fish alone or with one of the dogs. The urgency of catching something has quieted, and there are always details about a river, a flat, or a pond to keep my eyes occupied.

Studying birds, particularly the tall, thin-legged species that hunt the shallows by stepping and stopping, and moving and halting in concert with the reactions of their prey—just like the progress of my Brittanys in a wheat field—sums up my weakness for pointing dogs.

As a young dog Heather would not take her eyes off me while we were in the boat. With age—and like her master— her interests are less focused on the mundanity of casting and retrieving and more on the fish at the end of the line. Between strikes I watch the little dog staring at the shoreline, at birds and turtles. She is much like me and I am much like her. For years Heather never left my side. To walk or hunt or take a trip into town without her would have been inconceivable. Recently, if she loses sight of me during our walks, she is apt to lose her bearings and that frightens her. Consequently, when I leave the house these days, she looks at me and, unless I order her to follow, turns back and goes off to find my wife. For the obvious and inescapable reasons, her age and anxiety break my heart, so instead of walking we go fishing.

Between strikes I think about what to cook for supper and what book to read afterward. I embellish the looks of the girls I used to date and sometimes daydream about buying the winning lottery ticket. I never think about human poverty, disease, corruption, pain, or the fate of the planet when I'm on the water. I save that for the ugly places: strip malls, waiting rooms, drive-through restaurants, the inside of cars, cities, suburbs. I also talk to fish. Not fluid conversations, but expletives that punctuate the intensity of the action. A missed strike or a lost fish

warrants a "Shit!" A poor cast a "Damn!" A good fight "Bravo!"
A release "Salut!" The rest of the time I am silent, except for a
short reprimand when Heather leaps at the fish hanging from
the hook.

Because I am surrounded by sentient lives that subsist free
from man's ambitions and the boundaries of his pollution, a
more difficult subject for me to avoid assessing when fishing is
the treatment of animals by alleged sportsmen, the food industry, or the medical community. The pond is my sanctuary, a
hallowed place from where communion is taken and confession
expected, and where, because I am so close to the natural life
of its denizens, I often find myself dredging up all the terrible
things I have done to animals in my life. The thought of harm-
ing innocent beings has been abhorrent to me ever since I grew
out of my heritage, which encouraged killing for killing's sake.
Unfortunately that shift took a long time in coming, and even
now, while I entertain certain reservations, I continue to slice
behind the gills of a fish in order to remove its fillets and pluck
the birds that Heather retrieves in order to access their flesh.

These victims of civilization have always presented me with
a conundrum that creates a cognitive dissonance, a toggling
back and forth between the wonder and empathy for the king-
dom of nature and the desire to harvest for my own inherited
reasons. I rationalize my actions by repeating the aphorism,
"The price for living is death." But I know better. Too many
birds were shot and too many fish caught and abandoned on the
riverbank. The only reason to kill is to eat. May my conscience
punish me into grace.

With Heather observing, I cast at the vertical shoreline of
the dam and watch the worm sink out of sight. Everyone has
a preferred method of fishing a worm. Mine is simpler than
most: reel the worm slowly, just above the bottom. With every

revolution of the reel, I raise the rod and twitch the tip, then drop the rod back, parallel to the water. The worm moves forward a little, rises and falls in a column of water, then moves forward again toward the boat. The aim is to swim the lure as deliberately and slowly underwater as one can imagine a worm would swim. Since the take happens out of sight, there is nothing exciting about worm fishing except when a big fish jumps it out of the water. The take is usually subtle and often happens as the worm is falling; a line feeding away from the rod suggests that a bass is repositioning the bait. Most anglers have, at one point or another in their bass-fishing life, lost a double-digit-weight fish. Just as they do after missing an open shot at a ruffed grouse, they tend to remember the occasion. Usually a lost fish implies that the angler did not strike the bass hard enough, or by not applying the correct amount of pressure, allowed it to wrap the line around debris on the bottom.

My unhurried drift back to the pond house takes me past a grove of cypress trees that during a year of regular rainfall protects a small bay. Because of the drought the trees are growing on the bank, their roots uncovered by water. Pinched between two clouds, a ray of sunlight falls on their knees, smooth, clay-colored, knuckle-shaped protuberances growing close to their base. It was believed that these secondary trunks drew oxygen from the air and distributed it back to the tree. In fact no one has any idea of their purpose. At Southern fairs, local "artists" display plasma sculptures and lava lamps assembled on cypress knees, perhaps to fixate on while smoking a joint. In antiquity the cypress tree was a symbol of mourning. The Greeks and Romans believed the cypress was the first tree the dead would see when they reached the Underworld. In the past I have caught bass lingering next to cypress knees. Not today. It will take a foot of rain to raise the water far enough up their legs.

The first and second bass, both shorter than the seven-inch worms used as a lure, sail out of the water when I strike. Despite her years, Heather lunges at each fish as it comes over the side of the boat, displaying an exuberance that time has not yet tamed. There are too many bass of that weight class living in the pond, but because they are too small to eat, I cannot bring myself to cull them. My dog watches me toss them back.

The next two fish are each about fifteen inches long and will surrender perfect fillets. The first bass takes the worm a long way from the boat, almost to the shore. A few casts later I watch Heather cock her head to one side, and two seconds later I feel the strike.

I used to clean a hundred fish a year. As it is with anything physical repeated often enough, the assignment eventually becomes second nature. A sharp knife and orderly strokes get the job done. Now that I only dress a few fish a year, my knife work leaves much to be desired.

This day I place the bass fillets on the bottom of four squares of banana leaves and pour over them oyster sauce, mirin wine, and shaved fresh Thai peppers I grow in the herb garden. Folded into packages and grilled, the fish and sauce steam inside the banana leaves, which impart their own flavor to the fillets. Along with snow peas and green onions, it makes for a satisfying spring lunch.

The pond looks younger and less emotional than it did in March. The wind and pollen have subsided, and the branches of the cypress trees show a deeper shade of color. At night the male spring peepers sing. For days I've been seeing gangs of teenage bass team up on schools of threadfin shad in deep water. Most of the bass involved weigh about a pound, but that doesn't stop the raids from being fierce. During the attacks, the surface of the pond erupts, as if strafed by shrapnel. Expelled from the water, small, terrified shad scatter across the surface like silver coins. I look hard through binoculars, but I can't tell how often the bass are successful.

I fish for them by aiming the boat at the surface disturbances, guessing where the shad will emerge next. I use the four-pound test spinning rod and a three-inch Rapala lure. Each time the lure reaches the bait, I hook up. Generally the bass are small, but on every fifth or sixth cast something of size takes the lure deep. Heather sits alert and attentive on the bow next to me. The action is fast.

Since I hate treble hooks, and to facilitate releasing fish without ripping their faces off, I bend down the barbs. Having done so, I lose several bass that breach water. It doesn't bother me. I keep two fish to eat and return the others. The pond surrenders sixteen bass, the largest weighing four pounds. Some fish I cannot handle, which is why I use a light line. I don't see the point in fighting tarpon or billfish anymore, but I love to play bass.

The activity in the pond reminded me of a lunch in France ten years ago, a long thirty-two-course meal that started at one in the afternoon and ended seven and a half hours later. The motif was eighteenth-century French cuisine. After seven or eight courses—or approximately two hours' worth of eating— all fourteen diners took time out to stretch their legs and settle their stomachs by strolling in the gardens adjoining the dining room. By the third time-out, tired and sated, individual strollers and small groups of diners passed each other, burping and cutting small farts while commenting on the veracity of the food. The verbal analysis of each course rose in equal volume and predictability to the fullness of everyone's stomachs, just as my afternoon of chasing schools of harassed shad became progressively more mundane each time I caught a bass and released it.

I know of an eighty-four-year-old freshwater guide in south-west Florida who admits that he rarely hooks a bass that weighs over one pound. He still fishes every day, the way a kid would, for the joy of being on the water where the Seminole Indians live and the saw grass grows tall, where the white pelicans winter and the last of the Florida panthers mate. Mostly he fishes where those who favor malls don't visit.

"The greatest sight in fishing is the hole where the popping bug used to be," he is known to say.

A sight I hope to see each time I make a cast.

Fishing the pond in the spring is about looking at a clean body of water starting to bloom. At the start of the season, the cypress trees are still brown and almost bald. For months they've been the scarecrows of the pond. But by May, when the early growth on their arms matures, their branches develop long, narrow spirals of green leaves that will effectively shade the pond and cool the shallows until Christmas.

Already the quail whistle, and the doves fly in pairs. Early in the morning jakes and gobblers tumble out of their roosts in the branches of loblolly trees to the forest floor, each one looking for a hen that might be receptive to the magnificent display of its tail feathers.

Overnight the banks of the pond turned green, as if a paintbrush had been drawn across the mud. Because a heavy rain the day before had raised the water level a few inches, the bream fry had flocked an inch closer to the pond's original boundaries, miniature silver fish canting obliquely at first light.

On this first day in June, the pond is as smooth as the glass on the window of the pond house the blackbirds have taken to task—an error that leaves a few stupefied, some dead. The pond glistens, as if coated in tung oil. Adult bream dimple the surface. I pursue their rises from inside the boat and cast tiny yellow poppers with black tail feathers. Sitting down and casting close to the water is more personal than casting while standing. The white rubber legs of the poppers appear to row the bug back to the boat, an action scrutinized by Heather. If the cast is on par with the rise, a bream will take the popper. Going from one disturbance to the next takes time, but it is time well exploited.

I use a nine-foot, five-weight fly rod and six-pound test leader. The bigger bream bow the rod nicely. The biggest fish are called "titty" breams because they have to be held against one's chest in order to unhook them; they are broad enough to fillet. For lunch I cook them slowly in a cast-iron skillet with butter and a splash of hot sauce.

The season of heat in northern Florida begins in June. Vanguard thunderstorms take aim at the farm from the west. Otherwise the weather is tender, with gossamer colors and light winds that carry the sound of male quail singing. From inside the boat I imagine the shudder of bream spawning. A female bluegill, her anal vent swollen, twitches over the nest the male has readied for her to lay the eggs he will fertilize. The pond is orgasmic.

I brought a sandwich with me in the boat this morning and slipped slowly around the pond, close to shore, looking for mating frogs. The night before, the bullfrogs next to the pond house had roared their passion, sounding like a soccer stadium filled with vuvuzelas. It had rained before sunrise, the hum of water on the tin roof drowning the vocal expletives but not the ardor of my subjects. The pond had woken clean and lovely, like a thin-skinned girl at the end of her morning shower.

Over the years I've heard alligators, cicadas, owls, eagles, coyotes, and every so often a primal scream rise from the margins of the pond, sounds that herald the night, sounds that convey terror. Even sounds coming from small green frogs, amphibians typically eaten by larger frogs or birds with long hard beaks, snakes, or the always present alligator tick (a giant water beetle that inserts digestive saliva through its nostrum into the frog and sucks out its pureed remains).

Such observations rarely make good starters for social conversation, but I am old-fashioned and still dance to the high notes of a clarinet. Watching a pond evolve from minnows swimming circles around a yellow nightlight to eagles eating catfish is a waste of time, but that's how I've wasted my time for decades. Do I have regrets? A few, but they don't last. So I go on studying glossy ibises stalk frogs in the shallows, alligators swimming toward those disturbances that might signal food, the wind's authority over the surface of the water, and the golden light of the afternoon gathered on the face of the pond. I spend a great deal of my time looking out the window in front of my desk, a portrait of the sky with water below and pine trees in between: the cameo of my pond.

THE ISLANDS

By the end of the 1957 academic year, the headmaster of the boys-only French boarding school I had been attending made it clear to my parents that the odds of their son passing his baccalaureates were not worth betting on. With little ado I was flown across the Atlantic Ocean and enrolled into a coeducational boarding school in southern Florida where, it was presumed, the curriculum would be better tailored to my abilities. The student body in that school comprised fifty boys and seventy girls, and to my undying gratitude, twenty of those girls were gorgeous. Having just spent six years in an institution where gray clouds, cold rain, and corporal punishment (the act of caning was looked upon as a *métier*) were everyday occurrences, the move to Florida—the state of sunshine, tan lines, and sexual fantasies—is where I matured in all regards, except scholastically.

In the spring of my first year, Gilbert Drake, a student two years ahead of me, invited three of his friends to his father's newly acquired island located off the east end of Grand Bahamas Island. Deep Water Cay, which the Drake family had leased from the British Crown for ninety-nine years, was a narrow, one-mile-long spit of sand and coral that they would soon fashion into the best bonefishing camp in the Bahamas.

Over the months and years, Gil and his brother Tommy, as well as a succession of contractors, built the dock, generator house, lodge, guest quarters, private houses, and airstrip. I came and went as often as school and family obligations allowed and was generously provided with room and board, against what limited help I could provide with the jobs at hand. My mechanical skills did not match my unlimited enthusiasm, and it soon became apparent that my inclusion into the Drake family was at my friend's insistence. Of those years I possess more memories than it is possible to have lived, but I was young then and happy to fight windmills. Just as I now forgive my son for aggrandizing the memories of his youth, I forgive myself for making up a few of my own.

Before the lodge was built, we slept on the deck of the *Magic*, a wood-hulled thirty-six-foot Nova Scotia inboard, anchored in the current of the creek to confuse the mosquitoes. The island was short on amenities. For a long time frozen meat and poultry, fresh vegetables, fruit, and even blocks of ice were imported from Nassau. Gil and I would take turns running a skiff out to deep water on the far side of the reef and guiding the rusty sixty-foot supply ship through the natural and dynamited gaps and corridors in the coral leading to the lodge.

The *Magic* made half a dozen trips every season from South Florida across the Gulf Stream, carrying whatever was needed to build a fish camp. When I would join Gil for the return trip, we would leave West Palm Beach at dawn for the five-hour crossing to Freeport and point the bow of the *Magic* at the rising orange ball of sun that greeted us as we left the inlet. At times it was a long, quiet ride through the infinity of a young man's dream, a perfect adventure with danger hiding so far below the hull that it never rose to the immediacy of reality. But the Gulf Stream had a habit of turning without much warning from the calmest horizon to a fierce old fighter, gray and screaming and unpredictable because of the direction of its current and the wind that challenged it. As teenagers we felt we understood the sea and how to deal with its particular fickleness, which at any rate turned out to be effortless compared to the moods of the girls we dated.

After a few years into the building of the camp, Cessna 310s started using the coral-and-grass strip we had helped carve out of the mangroves, ferrying guests from the mainland to Freeport, and then from Customs to the island, where a tractor-pulled wagon greeted the plane. Taking off from Deep Water Cay with five full seats and the absurd amount of baggage that anglers, to this day, insist is de rigueur for catching a five-pound fish was an experience I qualified as a "holy shit" moment. More often than not, the second turbo-charged engine refused the hot start. So while the passengers sat in their seats at the end of the runway, staining their city clothes with sweat, the pilot would crank and crank on the engine until suddenly the plane bucked and howled into life. It was then a matter of applying full power while standing on the brakes and, when the 310 hunkered down and felt like it was going to disintegrate, letting go. Determined to meet the passengers' connecting flights in Florida, the pilot would lead the Cessna around a magnitude of rising clouds and rain showers, and through the ephemeral beauty of Southern rainbows. Predictably, turbulence was present to greet the plane every time it lifted off the ground.

In June the water was warm and clear and as yet unhindered by the proliferation of plankton that would bemuse the ocean during the oppressive days of August. Tremendous cloud formations reached up into the sky every afternoon. Late in the evenings they fell into the ocean in avalanches of rain and lightning and thunder. When it was dark and a mature storm chose the east end of Grand Bahamas Island to flex its authority, thunder rolled across the flats like the outbreak of war. The sounds of doom and disaster were such that I would crawl under my bed and lie facedown, holding a pillow over my head.

Gil and I, like most young men, enjoyed tempting fate. On any given day we would run one of the lodge's fourteen-foot bonefish skiffs out of sight of land to look for something, anything, out of the ordinary. We free dove in a thousand feet of blue water with miniature dolphin and sailfish under mats of Sargasso weeds. We chummed for sharks among schools of yellowfin tuna, and if the opportunity presented itself, we would run the skiff up the back side of funnel clouds and, from a football field away, watch the gray column of water vacuum the ocean into the sky. As often as we could, wet and caked in salt, we would hang on to the bow rope and ride the skiff like a horse, burning a trail of adventure across the greatest playing field on Earth: the sea.

Years passed before I would realize how lucky I had been to live on a Bahamian island in the late fifties and sixties, a time when phone calls to the mainland could take days, when the affairs of men were sealed with a handshake, and when the sport of saltwater fly fishing was in its infancy. We waded the flats without ever stepping in the footprints of another angler and dove in water so clear our shadows mottled the sand below us.

What were everyday events to us are to most people today inconceivable.

Gil and I defined adventure as anything unusual, memorable, or frightening; if the concept of beauty entered into the equation, all the better. Bonefishing did not meet any of the criteria, but being dragged through the water by a speared lemon shark met them all, as did floating on top of a thirty-foot reef at night without a light. Snorkeling across a shallow, white flat and being swallowed into a bottomless blue hole where snappers and amberjacks patrolled the entrance also qualified. In those underwater caverns the reef fish waltzed and the prismatic reflection of the sun on the walls of the cave wrapped itself around us as we descended into a honeycombed world of tunnels and corridors, at the bottom of which lay sleeping sharks.

Encouraged by the tide, there were days that we snorkeled over miles and miles of reefs, our masks wondrous magnifying glasses that exposed us to the green heads of moray eels, the wide pink lips of conches, the simple-minded advances of hog fish, the exaggerated eyes of Nassau groupers, and the occasional school of bonefish patiently waiting for a surge of water to carry them onto the flats. In deep water the bonefish were unassuming, bland, and undistinguished for being a single member of a school with no apparent purpose. In shallow water the subtle blending of mangrove colors on the fish's head and back, the milk white of its belly, and the breath of powder blue on the brow of its fins transformed bonefish from bottom-feeders to the speeding darlings of the angling community.

Gil and I, however, wanted fish that pulled, fish that jerked and tugged and dove and tried to rip the rod out of our hands, fish like marlins and sharks and tunas. We liked casting and dragging wooden plugs across the reef at night, plugs that under the light of the moon left iridescent contrails below the surface. We loved being startled by the slashing of water triggered by a barracuda sideswiping the lure, and the way the sea blew open when a grouper rushed the plug from below. In general, we looked for anything that put up a fight, with special emphasis on the unfamiliar.

Black-and-white photographs of us from those early days portray thin boys in black tank suits, all knees and elbows, awkward in demeanor and yet wearing the know-it-all smirk of youth; boys caught on film between adventures; boys who dove with fins and snorkels and speared everything in sight, including huge stingrays for the thrill of being dragged through the water by genetically old creatures destined to die before the day was over. On moonless nights we would test our manhood by standing at the end of the dock tethered to a rope tied to a hook and a big chunk of fish. The bait settled to the bottom a hundred feet out in the falling tide, and we waited for the inevitable shark to pick it up. The winner fought the shark back to the dock or followed it overboard; the loser cut himself loose before being dragged in.

We snorkeled the shallow reefs at night with cheap flashlights and broom handles and laughed hysterically when forced to fend off those small blacktip sharks interested in our fins. At night we dreamed of treasures and billfish and island girls, ever confident in what the new day would bring. Moray eels, rays, sharks, turtles, ducks, pigeons, and more were all there for the taking—for the sport, for the eating, and to fulfill a basic wish in men young and old to exert dominion over the natural world. Even though we eventually outgrew the need for killing, fishing was in our blood and would remain a part of our lives forever.

Our favorite time to be on the island was in the summer after the lodge was closed to the paying guests. It was then that the realization of every sporting dream was simply a matter of waking up. We left the dock each day after breakfast loaded with fishing tackle, diving gear, and sometimes a floating, collapsible shark cage made out of PVC tubing. We'd take refuge in the tubing while we photographed those sharks that got overly excited at the chum we bled into the water.

Days that started with the vaguest of plans were invariably amended and shaped to conform to the weather and the mood of the moment. Sometimes we shot flying fish from the bow of the skiff while running wide open over flat calm water, our targets reflecting the sunlight before spinning into the blue, out of sight. Other times we chummed the ocean for whitetip sharks, for the thrill of watching them rise out of the ocean and—more to the point—for the thrill of scaring ourselves senseless. In September we would hunt legions of crayfish marching single file in the lee of shallow coral reefs. We dove for them, gigged them, speared them, and ate them fresh, prepared by big, cheerful black women from McLeans Town, women who shuffled across the kitchen on heelless shoes and fussed at each other while they cooked. Superstitious women, who, at the mere mention of Condosha, the ghost of an African slave resurrected as a black dog that ate human flesh, would run off the cay terrorized and not return for days.

We lived in the present, because that's what young people do. When a frigate bird spread its black wings on the wind or a sailfish billed through a school of bait, when the creek took on the deep delight of greens and blues that swept in from the ocean, overflowing the flats with fresh water and food, we didn't define the moment as anything other than being alive. Dawn was a pale affair, a triumph of gradualism and the preface to a new adventure; nights were thick with the sounds of insects and crabs, the pleasures of exhaustion, and the wonder of dreams.

Imagine a mirror granting every wish, and then imagine looking into that mirror every single day.

It was at the long communal dinner tables inside the lodge that drinks and food found comfort in the soul of the paying guests. The anglers were for the most part middle-aged or older men and women, CEOs of active companies and retired business people from New England, guests with oddly pompous accents and money to spend. Overall good people, the guests dressed in universal khaki and brought with them high expectations. They had paid not only to fish but to be pampered, and in the evenings they confidently put forward their opinions on a variety of subjects.

Piscatorial geography, fishing lodges, boats, guides, lures, and of course politics were the foundations of the energetic diatribes that two or more martinis at cocktail hour drew out of the otherwise polite, conservative guests. Unbeknownst to them, they discussed the same subjects, asked the same questions, and put forward the same conclusions the anglers they had replaced had submitted the week before. After a day in the sun, a few drinks could turn these civil men and women into Jekyll and Hydes whose voices rose in proportion to the liquor they consumed, until half a dozen well-educated New Englanders were braying like donkeys.

Gil and I considered bonefish as undemanding to catch as a grunt, and we could not understand the nightly arguments related to the angling techniques applied to *Albula vulpes*. A spinning rod, a shrimp on the hook, and the wherewithal to land the bait inside a target the size of an armchair at sixty feet ensured a grab. Most of the guests who visited the island did not display that particular ability, however, and were delighted to release three or four fish a day. These caricatures of our future selves— old men with white hair, bony heads, and soft bodies—were to us of bleached hair and tanned feet cause for unending wonder.

The Deep Water Cay guides were from McLeans Town, one cay over from the lodge. With thick Bahamian accents they told and retold stories of their fathers sailing eighteen-foot dinghies south to Cuba for sugar and as far north as the Carolinas for liquor. They passed on the legends of treasure inside ships torn open on reefs during hurricanes, and of their whereabouts in relation to a spit of land, a depth of water, or a familiar cove. Dark men of varied ages, their ancestors had been abducted from West African villages, transported in chains across the ocean in the guts of slow-sailing vessels, and put to work by men who displayed more empathy for the life and death of their livestock than that of their slaves.

The children of those first slaves had survived the rhythms and rigors of life on the nonvolcanic island of Grand Bahamas—an island that rose out of the sea through the pressures of time, an island of simple beauty but lean of earth, vegetation, and compassion. In turn, their children now survived by harvesting bonefish with nets, crawfish with ticklers poles and gigs, reef fish with hand lines, and conchs out of small dinghies.

Some of the guides at Deep Water Cay were better than others, but they all wore radiant smiles bright with hope. They could see fish on the flats without the help of anything other than the eyes they were born with—the same eyes that would cloud over later in life from decades of staring through the reflection of sun on water—fine men with one foot in the past and the other in a tentative future as bonefish guides to wealthy men and women.

Regulars at Deep Water Cay included such famous anglers as Al McClane, Tom McNally, Joe Brooks, and other professionals who had introduced the sport of saltwater fly fishing to trout fishermen, men who had caught the strongest, wildest, and heaviest of all fish on fly. These men came back to the island year after year, in part because of their friendship with the owner, Gil's father, but also because they beheld fishing the flats as a pilgrimage and hooking bonefish on fly as a hallowed experience. We talked to them, these skilled men with tough hides and narrow eyes, learning from their stories, their egos, their lies. We saw in their faces our own mortality. But when it was all said and done, the difference was fundamental: We fished as children do, for pleasure, while these men fished for a living, and something about that forgave us in the eyes of God.

None of the lore, however, and none of the excitement or suspense that impelled so many of the guests to return each season to Deep Water Cay to stalk bonefish made sense to me until the day I tried to duplicate the flight of a shrimp with that of a fly.

Suddenly, what had been effortless grew in complexity, what had been instinctive felt awkward. Second nature no longer applied. From that first moment of holding a fly rod—that first tangle of line, those collapsed casts, the wind knots, and the first sting of a hook in my neck—bonefish magically transformed from an offshore trolling bait and appetizer baked and splayed on the bar at cocktail hour to a glowing target at once simple and yet as complex to decipher as a Japanese koan.

I still thought of the fish as forked-tailed speedsters, silver-bodied bottom-feeders, clowns that ate standing on their heads. But because casting a fly is all about angles and wind and presentation and focus, and because I wanted to master those demanding and unfamiliar mechanics, I learned to appreciate the fish as a nervous, gray target that changed directions on a whim and when troubled raced away at the speed of a falling star.

In 1964 my fishing world changed forever. Every fish species I had caught on regular tackle was now hunted on fly, from bonefish to sharks, from barracudas to tuna to sailfish to tarpon. I did not use a spinning or bait-casting reel or an offshore rod for the next thirty-five years.

TO CAST A FLY

Half a millennium ago someone had the idea to attach a measure of line to the end of a long cane pole and use its reach
to entice trout and salmon with crickets, flies, and grasshoppers. Rudimentary wooden reels with extra line encouraged
the ensuing generation of anglers to cast the fly beyond the tip
of the rod. Today the intention is the same, but the tackle has
evolved into light, strong, flexible carbon poles and reels with
drags that allow the backing to whisper after the fish smoothly,
effortlessly. In this century a ninety-foot cast, mostly useless for
fishing, is the modern measure of a man's fly-casting manhood.

The more accomplished casters understand that haste is
not a fundamental component of fly fishing. Staged within a
rhythm dictated by history and circumstances, the artificial
flight of a fly is meant to embody accuracy and temptation. A
slow cancelation of gravity, a cadenced motion that matches the
stroke of a scythe cutting wheat, the cast originates from an
old and beautiful tradition of men and women eager to entice
a fish using the feathers of a bird. The motion of the fly rod
is designed to stir life into the feathers, positioning them in a
horizontal runway in space where the fly takes flight and is sustained like a guitar note—once, twice, three times—as it gathers speed and direction until, finally released, it sails to a place
where fish swim.

I understood the mechanics of fly casting early, and within days I felt the weight of the line tugging at the tip of the rod and could effectually make a cast. But when it came to accuracy, particularly in windy conditions or while trying to present the fly at a target that moved of its own free will, months passed before I displayed any measure of control. I spent hours practicing on the beaches and golf courses of southern Florida and stripped innumerable fly lines across the blacktop on the street behind my house. I built wigs of monofilament, drove hookless flies into my face, and a few times, vexed beyond restraint, smacked fiberglass rods across the trunks of palm trees, the roofs of cars, and on cement walkways, each time triggering a pleasing shattering of fiberglass. I had a temper.

Gil taught me early on how to speed up the development of the fly line by double hauling on the backcast and again as the fly was moving forward. For a while, knowing when to pull and when to ease the tension on the line was like trying to rub my stomach while patting my head, but eventually the motion became natural. The subtleties required to master the particulars of catching fish, however, would take longer.

Time spent on the water gradually taught me the nuances that shape a good fly fisherman. I learned that an uninspired backcast guaranteed an equally weak delivery. Creeping the butt of the rod forward before releasing the fly line dampens the power stroke, as does collapsing the wrist inward at the end of the cast. A following wind begs for a low backcast and high finish; the opposite holds true in a headwind. A right-hand wind that steers the fly at one's head is overcome by learning to cast and duck or, more elegantly, taking the backcast over the left shoulder. Roll casts are used to avoid vegetation menacing the flight of the fly, and when confronted with shoreline growth, one learns the sidearm cast, parallel and close to the water.

It took ten years before I was comfortable with the arcane understanding of angles, wind, the vagaries of fish, and in the case of flats fishing—which is what I loved most—the all-important speed and direction of the skiff I was fishing from. Back then I could not understand why the techniques of fly casting were so elusive, but I now realize that it takes a decade to do most things well.

When the variables finally came together, I no longer thought about the mechanics of the cast. The fly line developed and stretched behind me because I expected it to, allowing me to focus on the whimsical nature of the target. The casts were mellifluous and silky on calm days, determined on windy days. I no longer felt a discord between my hands and my desires; the rod simply ceased to exist.

SUMMER POND

In July the bass fishing in northern Florida slows down. The water temperature is uncomfortably warm, and the fish congregate in deep water during the heat of the day. Deep jigging or night fishing works best. Since I find that wine and tying knots, wine and launching boats, and wine and casting do not make for a comfortable union, and since I enjoy the "mother's milk" of my father's land, I choose wine over fish.

Early one morning I see a low-slung 1989 Pontiac containing four ladies from the nearby village of Midway—granddaughters of the local black families who began their lives as slaves—drive up on the dam. An armful of whip-thin cane fishing poles, twelve feet long, stretches out through a side back window. As the ladies spill out of the car, their broad straw hats bejeweled with colorful ribbons catch the breeze. Pale green, bright yellow, and purple dresses decorate the weir. From a distance the dam has bloomed.

The driver, the younger brother of one of the women, unloads the gear and positions four aluminum folding armchairs at close intervals on the dam. He sets dog food cans full of crickets and white plastic buckets to accommodate the bream next to each seat. Bream fishing in the South, among all social sets, calls for social interaction. In this case the brother ignores the incessant clucking surrounding him. When the ladies are comfortably seated in their chairs, he hands them ziplock bags filled with boiled peanuts and backs his car off the dam, parks it out of earshot, and finds his pleasure: fishing for catfish miles from the rumpus of daily work and half a football field away from the chatter of his women.

The average age of the ladies is seventy-five. All are descendants of "freed" cotton pickers who endured the ignominy of Reconstruction and work in the northern Florida turpentine camps. It was their children—the ladies fishing on my dam and others of that generation—who performed the back-breaking farm labor required to tend the shade tobacco, a crop that, along with the access to bargain rate Coca-Cola shares, afforded the affluent families of Gadsden County bragging "rights" about living in the wealthiest county in Florida. Shade tobacco's handsome russet-colored tobacco leaves were destined to become the outside wrappers of prime Cuban cigars. Grown under row after row of fifteen-foot-wide white-and-yellow muslin awnings that stretched the length of the tobacco fields—shaded from the harmful effects of direct sunlight—weeded, watered, and well tended, the tobacco seeds grew into broad, beautiful, near-perfect plants.

Once picked, the leaves were dried in long, well-ventilated barns, high structures fashioned out of yellow pine. During the first week, the workers built fires to hasten the drying process. After that the tobacco was left to cure for thirty days.

Half a century later, the odor of prime tobacco clings to the walls of the remaining barns, while the ladies who helped tend to the tobacco leaves for shipping were relaxed on the dam, talking, laughing, and all the while paying close attention to the bobbing of each other's corks.

The first boiled green peanuts of the year have appeared on the markets. My fisherwomen know just how to scald them. The meat comes out of the shell not too hard, not too soggy, firm and just salty enough to call for a glass of ice tea—or in my case, cold beer.

I see some of these ladies at the open market on Saturdays in Tallahassee, sitting on low stools chucking peas directly into their dresses, content to be resting. Now old women falling to bone, they grew up poor and worked and prayed hard all their lives with the hope that their progeny would enjoy some measure of economic and social equity. In this part of the United States, at least, such dreams are a generation, probably two, premature.

The ladies on the dam are there to catch bluegills and shell-crackers. They tell me the smaller the bream, the better it "eats." I had read in the paper a month ago that one of the ladies had died. Today it is the remaining four friends who have come to the dam.

"We got no right to holler about such things as growing old and dying," one of them says to me, her face grayed by work and poverty. The other two ladies nod. One of the bobbers sinks, and presently a bream dangles in the air among squeals of delight.

The lucky lady looks up at her catch and says, "Fish, you can get accustomed to hanging if you hang long enough!" More laughter.

Frisky, the bluegills carry an olive-colored tinge, the males darkening to a rich purple hue during mating season. Shell-crackers, known as redear sunfish, sport a small characteristic red spot at the rear of their gill plate. Cleaned and dropped whole into hot oil, the tiny bream crisp up to a one-bite crunch.

The brother will accept anything that takes his bait, but on this day he is geared to catch channel cats. I had seen him two or three days earlier seeding the water with a special mix of catfish food that stunk to high heaven. Today the fish were gathering on the chum, and I watch him pull a mess of three-pound cats out onto the bank.

A short, heavyset man in baggie jeans, Alistair wears a white T-shirt and black baseball cap. While fishing he leans forward, his thumbs hooked in his front pockets, his attention focused on the meanderings of his bobbers. His wallet sags low in the rear pocket of his jeans.

We had talked one morning on the dam back in January, after he had bought a dozen golden minnows from the bait store because Bill Poppell had told him that the crappie were bedding and biting.

"To be honest," he'd said in a high-pitched Southern voice, "I don't like catching no fish that's on the nest. Too much like kicking a pregnant woman."

"You ever kicked a woman?" I asked him.

He smiled, his front teeth banded in gold.

"Once," he said. "In the butt. But not hard," he added for my benefit.

"What did she do?"

"Ate my supper."

The pond is hot and loud and teeming with wildlife and complaints in August. The community of fish has moved to deep water, and the frog chorus is booming and more vibrant with desire than it was a month ago. After dark a big fellow under the house sounds like a Brahma bull. A first coat of golden paint envelops the cypress trees. On slow fishing days I switch my attention to insects. Today it's the dragonfly, a creature that in fossil form had a three-foot wingspan. With compound eyes, dragonflies can see almost 360 degrees. They can fly as fast as thirty miles an hour, and in the nymph stage they spear and eat tiny fish. It is easy to imagine dragonflies the size of dachshunds on the prowl.

Ponds all across the South, from Georgia to Louisiana, boil over in summer. Rain impels algae in the water to propagate.

Mud rises to the surface. The fish won't bite. Sour water silences the pond. Nothing moves. Fish and fowl wait for the mud to settle back to where it belongs. At the feeder under the oak tree next to the cabin, the blackbirds rest, their beaks open, seeking the succor of shade. Red as paint, the cardinals stand guard. The thermometer rises to ninety-seven degrees. The air is thick and buttery, the heat merciless. There is little to do and no great urgency to get it done.

In early September I sense an edgy energy in the feeding and flight patterns of those birds preparing to migrate. There is renewed bickering at the feeder. A jamboree of blue-wing teal carve furrow patterns in the wind, veering and pitching, looking for a safe place to overnight on their travels south. As the sun sets I watch a dozen hunched silhouettes swim into the shadows—my first puddle duck visitors of the year.

The alligator population that checks in and out of the pond every spring and summer has been pared down to one, a small female living under the dock next to the pond house. She has taken to eating the fish food I offer the bream. In position when I walk out on the dock, she guards the closest half dozen floating pellets from the nearby bream and turtles, striking at them when they get too close. After some reptilian consideration, she finally nudges the round brown balls into her mouth with the tips of her teeth. The small specimen—I call her Olga—was born in the pond, and she already manifests the supple nonchalance of an alligator swimming.

How do I know she is a female alligator, since all I see is the surface of her spade-shaped head and three feet of corrugated lizard body? Because I decided a long time ago that if I cannot accurately identify the gender of a species by its markings, size, or colors, I catalog it as a female of the species. It is one of those "naturalist" quirks that prompt scientists to utter derogatory remarks about subjective conclusions, anecdotal evidence, and, the worst sin of all, anthropomorphism.

More than half a century ago, Roderick L. Haig-Brown wrote in *A River Never Sleeps,* "The North American is probably the luckiest fisherman in the world. He can range a whole continent of lakes and streams and up and down along the shores of two oceans; he can catch salmon and trout and char, bass and pike and muskie, sailfish and tarpon and tuna, and seldom enough run up against man or sign that seeks to bar his way."

Such is no longer the case.

The North American angler can still "catch salmon and trout and char, bass and pike and muskie, sailfish and tarpon and tuna," but if he doesn't own a private pond or stretch of river, a boat or permission to fish, he will run up against signs and men who will bar his way—or at the very least make the experience of fishing a public event rather than a pastime that has historically been private.

BALD EAGLE

Late in the afternoon the breeze is calm and the temperature has climbed to eighty-eight degrees. The surface of the pond is pockmarked with indications of fish rising. An early summer lethargy has settled over northern Florida, a precursor to the apathy of late August.

A disturbance in the sky across the pond catches my eye. One of the young ospreys, also known as fish hawks, hovers two hundred feet above the far shoreline that climbs into a shallow hill. From the porch of the cabin I hear its shrieks of displeasure. The young fish hawks have been on their own since their parents weaned them a fortnight ago. I grade the evolution of their fishing talents daily. Today the young osprey is beating his wings and screaming, its beak wide open. Something is causing the uproar.

I look for its sibling, but instead I see an immature bald eagle drop off its perch in a pine tree. It makes a half-hearted pass at the hawk, not to make contact but rather as a means to relocate the competition. The osprey flies to a nearby oak tree and watches the young eagle, whose neck and feathers are mottled brown and white. In the hierarchy of eagles it is a teenager; nevertheless its size is daunting to the likes of a four-month-old hawk. The eagle does not hover but flies out in the direction of whatever the osprey was watching; it makes a wide, open-wing turn and glides over the water.

As a kid I often watched bald eagles pluck mullet out of salt water in southern Florida. Later, with the advent of DDT, the birds came close to extinction, and for years I saw none. Now that the pesticide is outlawed, the eagles have made a comeback, and it is, as always, a delight to witness the exactitude with which they sweep over the surface, unfurl their talons, and, hardly troubling the surface, cleanly pick a meal up and out of its element. Usually the elegant coordination of the strike-then-flight offers a seamless resolution, with the eagle departing with dinner in its grasp. This simple act of power and precision is one I never tire of watching.

This time, as keen as the bald eagle is on stealing from the osprey, he underestimates the size of his quarry. When the bird's yellow talons sink into the catfish, the eagle instantly flips ass over kettle, landing headfirst into the pond. A surface battle ensues, with the osprey, back in flight, sixty feet above the commotion, watching. The eagle, his grip still secure, has righted himself. He swivels his head back and forth, looking for danger in the shape of an alligator or me. Slowly, using one wing as an oar, the eagle rows the fish closer to shore. With one talon on the catfish's neck, the other reaches for land.

The fight lasts three or four minutes, and when at last the duo reaches the bank, the bird is exhausted and gasps for air like the crows I see on summer days besieged by the heat. The eagle pays no attention to the fish hawk, who gives up and flies back to its nest. Rested, the bald eagle shakes himself clean and loses a few white down feathers in the process. With one talon holding on to the four-pound channel cat, the bird awkwardly drags the fish three feet up on the bank.

I watch this drama unfold from where I am rocking in the chair on the porch of the cabin. Observing such scenes is my excuse for being a slow writer. I keep the binoculars on the eagle and watch him bite into the anal cavity of his victim. After each rip of flesh, the big bird looks around for competition and danger with a measure of guts hanging from his beak. Pond turtles patrol the shoreline, and I understand why the eagle moved his dining experience up from the water's edge. Eventually a dozen turtles line up in the shallows below the bird's feet with their heads out of the water, ready to take charge when the eagle has had his fill.

I climb into the boat for a closer look, but by the time I reach the scene, the eagle has moved to a nearby pine limb. Eventually he will fly to a log in the pond and wash himself by repeatedly dropping his head underwater, raising it to the sky, and shaking his feathers like a dog. The turtles drop under the surface before the boat reaches the shore. On the bank all that is left are the ribs, head, and whiskers of the catfish.

The next morning the carcass is gone. The shoreline is cleaned by the ants. It is as though nothing happened.

There appears to be less mourning in the fish world than even in the world of Scottish puffins. If tears were shed for the catfish, they were shed underwater.

THE LIVING POND

When I bought the farm in 1990, the bass in the original pond were poor, their heads disproportionally large in comparison to their bodies, a phenomenon that infers acts of cannibalism. The bass had little to eat except themselves. To remedy the situation we had to provide more food and cover. Bill Poppell cut trees, fastened branches together, tied them to cement blocks, and pushed them into the excavated pond, while Charlie and his dozer, Gertie, worked on raising the dam. Once the tropical depression filled the hole with water, I bought gambusia, bream, shiners, and threadfin shad as additional nourishment for the starving bass population. In turn, the bait fish were fed floating, high-protein catfish food to promote growth and heighten their libido.

The pond harbors largemouth bass, speckled perch, channel catfish, and bream, which loosely includes bluegills, warmouths, stumpknockers, shellcrackers, and chinquapins, to name some of the more colorful synonyms. Speckled perch are locally known as crappie but carry the moniker of sac-a-lait (bag of milk) in Louisiana, specks in Michigan, and calico bass in New England.

Excluding the catfish, all belong to the sunfish family, species of fish that have entertained men and women of every age since before the first pilgrim landed. The most important

member of that family is the largemouth bass, the biggest specimen of the black bass species. Originally found east of the Mississippi, largemouths, hardy and sought after by anglers, were
introduced throughout the United States and around the world
before the turn of the last century.

The record largemouth bass was caught in Montgom-
ery Lake outside of Lumber City, Georgia, in 1932 during the
Depression by twenty-year-old farmer George W. Perry, who
was fishing to feed his family. His catch tipped the scale at 22.4
pounds. After weighing the fish at the local post office, the
Perry family ate it. The record still stands, with an addendum.
In 2009 thirty-two-year-old Manabu Kurita caught a bass on
Lake Biwa, one of Japan's largest bodies of water, using a live
bluegill. Manabu's bass weighed 22.5 pounds, one ounce more
than Perry's fish. The two fish weighing virtually the same, the
record is deemed tied.

To a saltwater angler, the shape and style of a bass resemble
that of a grouper. Both fish are stocky and built to fight. Discounting the fact that the bass will jump out of the water in an
effort to dislodge the hook, and the grouper won't, both fish
show speed over short distances and are equipped with gaping
mouths with which they aspirate their victims. Bass and grouper like to ambush their prey, bass from behind submerged logs
and structures, groupers from behind heads of coral. They are
both deepwater fish that foray occasionally into shallow water
to feed. On the surface, the open-mouth strike from either fish
is greedy and slings a substantial amount of water into the air.

The eyes of a largemouth bass are golden brown, its back
and head light to dark olive green (depending on the surrounding habitat) fading to a dirty white throat and belly. Its mouth
stretches to a point beyond its eyes and famously opens wider
than that of any other freshwater fish. Bass have good hearing
(tiny, well-developed internal bones pick up small sounds, such

as the clicking of a crawfish claw) and a reasonably good sense of smell. In clear water, bass can see up to thirty feet. Dark lateral stripes bisect both flanks and run from the fish's mouth through its eyes and down to its tail. These stripes are dark-colored pores that contain nerve endings that allow the fish to distinguish vibrations in the water.

That first year, once the rains came and the new pond was full of water, local friends who routinely fished Lake Jackson and the surrounding private ponds in the county kept the live bass they caught in aerated coolers and released them surreptitiously into my pond. I am not convinced of the legality of such a caper, but transporting bass from one pond to another is as foreseeable an occurrence in this part of the country as is the hundred-dollar handshake between a booster and a promising college football recruit. The new pond residents all weighed over three pounds, with some twice that size.

The swamp section of the pond represents a little more than one third of the total surface area, or nine acres of four-feet-deep shallows, a mirror of water that fills up with weeds during the summer and transforms itself into magnificent breeding grounds. For the next decade the bass grew and the baitfish multiplied. Catching a ten-pound fish was possible any time of the year, and a healthy number of double-digit bass were caught, some on a fly.

The simple twelve-foot boat I use on the pond is lighter and shorter than the skiffs I used to run in the Florida Keys in my thirties—years during which fly fishing distracted me to the exclusion of most everything else—fulfils my less nomadic needs, and is appropriate to the size of the pond. Three decades after fighting full-size pelagic fish, I've returned to a childlike wish for the simple pull of a bream or bass against whatever rod is suited to make that happen often. I cast artificial crank baits, rubber worms, and top-water plugs using a light spinning rod and a reel filled with either four- or six-pound test monofila-ment. A three-pound bass performs admirably on four-pound test monofilament, less so on the twenty-pound braided Dacron I have used while live-baiting on Lake Jackson, when fighting a bass was as challenging as surfing a dinner plate to the boat.

The sport of bass fishing was advanced through the stocking of largemouth and smallmouth bass throughout the country. Largemouth were moved to warm southern lakes and ponds; their smallmouth counterparts were transferred to north-ern lakes and rivers. By the 1800s wealthy anglers fly-fished for trout and salmon, while sustenance anglers—the work-ing class—fished for black bass using cane poles and live bait. The first artificial lure, which appeared in the mid-nineteenth century, was a variation of a salmon fly and a spinner that was used on fixed spool-casting reels and rods. Plugs, or floating wooden lures, were introduced around 1900, plastic worms soon after the Second World War. Spinning rods gained status in the 1950s. Sixty years later, the sport now boasts thirty mil-lion adherents and is a multi-billion-dollar industry. The per-son who breaks the sacred Perry/Kurita world largemouth bass record will win angling's mega ball lottery.

I keep a metal trash can full of catfish food on the porch of the cabin, out of which I feed the bream and shiners that reside under the dock. When I show up in the morning carrying the groceries, I'm greeted by an assembly of chaotic expectant fish. For a moment each morning I am the bishop of the pond, dis-pensing the Host to my congregation.

A cup full of pellets thrown into the water promotes a fish melee at the surface. The bass that otherwise spend their time in the shade of the dock are tipped off by the fracas and tear though the multitude of feeding bream and shiners, scattering them in terror. Smaller bass defer to the larger fish by granting them the first and best angles of attack. That said, in all the years I've fed bream off the dock, I've never seen a bass catch its prey. Once I watched a cripple surface from such an encounter, but the bass left it there, tilting from side to side, until a turtle clamped down on its anal fins and dragged it to the bottom.

In addition to my handouts, four solar-powered fish feeders spaced at intervals on the edge of the pond sling food into the water for two seconds, three times a day. Fatter bream are better equipped to breed and lay eggs and therefore more efficient at reproducing; the fry that emerge from this planned-parenting program equate to fatter, healthier bass.

One dark spring night a few years ago, I dropped a fancy underwater lightbulb tied to a weight over the end of the dock and turned it on. The globe of light rising from the bottom of the pond cast a yellow stain on the surface and attracted hundreds of fry, thin as pins that shifted in seeming unison, swimming quickly here and there with no real pattern for as long as I could stand watching them. Once in a while a yearling bass made a run through the lit-up water, but unlike the bats that successfully demoralized the mosquitoes outside the cabin's screen doors, the bass spent their youthful energy with no immediate evidence of success.

When I turned on the light three months later, the bulb beckoned through the denser, warmer water of summer. Scores of fry fed on plankton excited by the light. Out of the shadows young bass took turns exercising their skills, with the same outcome their contemporaries had displayed in the spring. It was a discouraging sight to witness, particularly since the fry were so small the young bass would have had to eat a plateful to make it worth their energy.

It wasn't until I moved to North Florida that I learned to appreciate largemouth bass as a culinary staple. With comparably little pollution in our ponds and lakes, and a lack of cow dung in the waters, our bass are firm and white, flaky and delicious. The same fish living in the canals of South Florida are soft fleshed and deliver the incongruous flavor of dead water. In northern Florida, where most foods are deep-fried, contrary to culinary customs way up north where the Yankees live, frying is akin to religion. A sinner may not go to hell for straying from the seventh commandment, but he will certainly end up down there for not minding his grease.

Ten years after the original ten-acre pond was transformed into a twenty-seven-acre lake, the numbers of big bass hooked or caught dropped. Instead of holding its own or improving, the pond began to decline in productivity. As usual, that prompted the standard blame game and finger pointing.

"So-and-so is keeping all the big bass."

"That person is filling coolers full of fish when you are out of town."

I listened to these and similar comments for a couple of years, closed the fishing down to all but a few trusted friends, and, when those measures did not realign the pond's equilibrium, called Charles Mesing, a friend who works for the Florida Fish and Wildlife Service. I've known Charles Mesing for twenty years, and this was not the first time he had electrofished the pond. It turns out, however, that I should have been seeing more of him over the years.

A big man with a wry sense of humor, Charles sports a circular beard that loops around his mouth—the kind of beard I have always thought would be uncomfortable to the woman being kissed. When not on the clock working for Fish and Wildlife, he runs a pond-management consulting business, dispensing advice on a pond's water quality, the legal stocking of fish, and the eradication of invasive aquatic plants such as hyacinths and hydrilla.

I met Charles and his partner Andy at the ramp early one June morning. His boat was an eighteen-foot Carolina skiff equipped with a pair of ten-foot fiberglass poles that, when in place, extended on either side of the bow like cockroach antennas. At the end of the poles, thick braided wires resembling dreadlocks hung from a metal hoop and dropped three feet into the water. An onboard generator drove a specific amount of voltage and amperage through the hollow poles to the thick metal wires. The setup reminded me of my youth in France when I would plug a 220V current into the moat and get into all sorts of trouble.

The modus operandi was to take a sample of fish—largemouth bass, bluegills, shellcrackers, shiners, and threadfin shad—along the shoreline over a set period of time, noting the numbers caught, their length, weight, and variety.

Andy maneuvered the boat along the shoreline with the generator working, while I stood next to Charles in the bow watching him chase fish with an oversize net. The bigger specimens would suddenly appear out of the darkness, canting from one side to the other, disappearing with a tail kick if Mesing wasn't quick with the net. In places of cover the fry rose off the floor of the pond like flocks of butterflies. The fish Charles caught were turned upside down into an aerated fifty-gallon drum filled with pond water to which he had added clove oil ethanol, a mild sedative to keep his temporary inmates quiet.

Andy stopped the boat and turned off the generator every fifteen minutes for a fish count. Charles retrieved the fish one by one from the barrel, announced the species, measured it, weighed it, and read the numbers back to Andy. After that, he turned the fish back into the pond where, stunned by the experience, it wallowed close to the boat before scrambling off none the worse for wear.

Those fish that made it into the barrel represented only a small percentage of the overall population of the pond. Had the survey been taken in late February or March and not in the summer, the bass would have been on their beds and more fish, especially the larger females, would have been netted, increasing the accuracy of the numbers. But I knew that Mesing, having taken similar surveys every month for twenty years, would have no difficulty in shaping a graph that—for better or worse—would give me a convincing report on the condition of the pond.

In one hour of shocking, 62 bass, 117 bluegills, and 34 shellcrackers made it in and out of the barrel. The weight of the bass averaged one pound, with only two fish of any size measured: a four-pound bass that sported an overly large head and narrow body, and a seven-and-a-half-pound fish, twenty-five inches long. Even taking into consideration that she had spawned back in February, the bigger of the two bass should have weighed ten pounds. It was obvious that she wasn't getting the food she needed to regain the weight she'd lost laying her eggs.

As we motored along, Charles pointed to the obvious lack of vegetation in and around the pond, caused in part by the drought but also because the trees we had pushed into the water twenty years earlier to provide cover had either wizened or rotted into the mud. The fry had few places to hide and grow, which meant they were being consumed while too small to give a growing bass a decent meal. Charles pointed at schools of tiny terrified bream huddled by the thousands in two inches of water next to the banks where the bass could not reach them. The numbers confirmed what I suspected: My pond was neither healthy nor sustainable.

Two basic things were needed: adequate cover for the small fish to hide in and grow and the removal every year of one hundred or so small bass that successfully competed with the big fish. Mesing suggested adding a few dozen gizzard shad to the pond in the winter before they spawned. Gizzard shad grow into larger meals than the threadfins, and their addition would bring diversity to the menu. His report also encouraged me to cut and push older, malformed hardwoods (the brushwood of pine trees disintegrates in one year) into the pond while the water was low, offering cover for the next five years.

Fish, like dogs, birds, and humans, are mortal. About 30 percent of bass die of natural causes, including predation, every year. Since bass grow at an optimal rate of one pound a year, a ten pounder is a survivor. At more than twice that weight, George Perry's and Manabu Kurita's fish were ogres. That we did not find many big bass in the pond made me reflect on my role in the health and well-being of the small body of water I had encouraged into being. I followed Charles's advice.

Although the trees we cut next to the pond over the next few weeks were dying or malformed, the process was, as always, a shock for me. To watch a tree being timbered, knowing that I would never see it again, engendered a sense of loss similar to that experienced upon meeting a close pal after a few years of separation and realizing that his dogs have grown old. Worse was the realization that, much like a friend or relative, the tree I would no longer see would be forgotten with the passage of time.

THE DROUGHT

When the sound of rain is absent for months at a time and the results are visible in the shape of failing crops, dying trees, and wizened ponds, and when sinkholes unlock daily and the land lies dispirited in the face of the drought, it is easy to mistake the sound of air-conditioning for rain. I have often hastened to the window only to be disheartened by the sight of small, spirited clouds and blue sky.

This year my pond is dreadfully low. It looks like a small prairie. By "Miss American Pie" standards, the levy is dry.

Three miles to the southeast of the pond, Lake Jackson—historically the most famous bass lake in the state of Florida—is drying up as well. At four thousand acres, a little more than eight square miles, it has always been too small to interest anglers on the professional bass tournament circuit. But when I was growing up in South Florida, the mention of Lake Jackson—shallow, rich, and full of bait—implied trophy bass just as the island of Bimini in the Bahamas evoked images of bluefin tuna hanging from overhead beams at the end of the docks, white weight numbers painted on their flanks.

Florida is known for its sinkholes, cavities (pockets of air) in the earth created over hundreds of thousands of years by erosion. When water fills the cavity, outward pressure on the walls supports them. But when the water table drops, exposing a hollow limestone chamber, the walls crumble and collapse. Anything above the hole is sucked down into it. Sinkholes swallow water, cars, trailer parks, and retirees with the regularity of a flushing toilet.

Lake Jackson has two major sinkholes. In times of severe drought, Porter Sink in the south section and Lime Sink in the northwest quadrant crack open and, over a period of weeks, suck all the water out of the reservoir.

After a rain-free summer in 1999, the water level of Lake Jackson began to fall dramatically. By February of the following year, the Fish and Wildlife Service allowed unrestricted fishing—no slots, no limits. For weeks families armed with rods and reels, nets and gigs drove their trucks and buggies to the few remaining pockets that held water and collectively stacked thousands of bass into their ice chests. I drove across the dry lake bed in March to examine Lime Sink firsthand. Two dozen cars were parked on the edge of the fifteen-foot-deep depression that bordered the actual hole. Fold-up chairs, picnic tables, coolers, pop cans, boom boxes, cane poles, bobbers, and every imaginable plaything that complements mass entertainment were on display. Men, women, and children were busy fishing the last gallons of lake water.

Hundreds of bass and bream swam in the basketball court–size depression housing the thirty-foot-wide sink, but they weren't biting. Opinions were put forward and argued between swallows of Coca-Cola and Bud Light. A fat woman who farted at an inopportune moment in the course of a disagreement giggled. The urban legend of the scuba diver in a lake who'd been scooped up by a flying boat during a firefight in California was debated. One fellow tied a large weighted treble hook to his line and cast it at the bewildered fish with the idea of snagging one, a common enough practice but one that was shouted down by all those who had not considered it first.

By the very end of the drawdown, the sink had sucked untold thousands of fish deep into the aquifer to live, perhaps as albinos, maybe blind, and certainly never to be seen again.

A week later the lake was dry. Dune buggies and trucks and motorcycles bearing Confederate flags took to the hardened mud.

When the lake dried up, Leon County and the City of Tallahassee, with the help of other government agencies and volunteers, initiated a cleanup. Over the next year a million and a half cubic yards of mud—one hundred thousand dump-truck loads—were removed from the southern end of the lake, the section most affected by human detritus. To their credit, many residents fanned out, on foot, over hundreds of accessible acres and removed the bottles and cans and plastic that herald our civilization. Others prayed.

"We churched it up a lot," one lady said to me, proud that her god had willed the residents to clean up their lake.

Two years later, in August 2001, tropical storm Barry dropped twelve inches of rain on Tallahassee and returned water to Lake Jackson. The following year the Fish and Wildlife Service reintroduced thousands of pounds of bass fry to the lake, and with cleaner water in which to proliferate, the fish population multiplied exponentially for ten years.

Unfortunately, just as the bass fishing was making a comeback to its former glory, a hard drought returned in 2010 and Porter Sink cracked open again. Five feet of water disappeared from the lake in so many days. Alligators returned to feasting on the bass confined to the shrinking pools, which in time further evaporated, exposing a fish smorgasbord to the rats and raccoons.

Historically Florida has depended on tropical activities to rejuvenate and keep its territory hydrated. Without rain, lakes shrivel, sinkholes yawn, and the state burns. Gadsden County averages fifty-eight inches of rainwater a year. In the two years following 2010, the county found itself positioned forty inches behind the eight ball. As the waters inexorably recede, the conundrum lies in wishing for a tropical storm without inviting a hurricane. The added despair for people of a certain age is not only the wait for enough water to fill the lake, but the fact that it takes a decade to breed a new generation of bass worth fishing for.

I read that when the drought of the 1930s finally broke over the plains, it rained mud until the topsoil was washed out of the clouds. Improbable at this latitude, but fires follow droughts and crops die and people suffer. And yet there are those who refuse to question the cause or consequences of the speed at which the planet is warming, the significance of melting glaciers, and the disappearance of species both large and small.

Dreams of rain take shape as altars of clouds. Heat leans on the land and transfers half an inch of evaporated water a day back to the heavens. With Lake Jackson draining and the shallow end of my pond looking like a ditch in Kansas, Bill and I talk about tropical depressions as if discussing a missing family member.

I watch the Weather Channel with enthusiasm and unfounded optimism bordering on the pathological, imploring rains closer and cursing when it doesn't happen. My wife, a lay Carmelite nun, points out that badmouthing her god has not proven to be an effective strategy. She may have a point.

Since this latest drought started, almost three years ago, logs have seemingly climbed four feet out of the mud in the shallow end of the pond and into view as fossilized arms. Except for random pockets, the water is only four inches deep. Grasses emerging from the mud stand tall, and minnows dance in ever smaller puddles of water. Trapped bream swim in place, their fins out of water that will soon be gone. Herons and egrets eagerly take up station. They have specifics to dine on.

It is a tragedy of some proportion, but since the pond is more than twenty feet deep next to the dam, the fish are not concerned. Turtles and the occasional alligator use the revealed rubble as basking stations on which to warm their blood. The shallow end of the pond has assumed an uncertain beauty of its own. Once the water table rises—and it will—the grass will provide food and cover for the fish to lay eggs and their off-spring to thrive.

In the pond's main body, the bream and shiner fry that typically hatch on warm afternoons have been forced outward from the safety of the swath of vegetation that usually lines the shorelines. Revealed in open water, the minnows find safety in numbers. The surface disturbances of a thousand small fish excite the herons, and belted kingfishers jostle with each other for the right to strafe. In slightly deeper water, bass patrol the outflow. From time to time the pond erupts in short, violent conflicts. Once or twice a day an eagle confiscates a fish.

The pond has no opinions on the drought that surrounds it. All she wants is to be a pond, a mother to her tenants. She is flexible to the weather and temperature, and as long as she holds a certain depth she will remain a pond. However, if, over time—decades, even centuries—and for whatever reason, the water leaves her basin, the aquatic plants will fall to the bottom and, offering their nutrients back to the soil, encourage emergent plants such as cattails, reeds, and sedges to replace them. The pond will become a swamp, and without the rain to fill her back up, she will lose more water through evaporation and eventually metamorphose into a marsh, then a bog. Finally the pond will become a wet meadow, a propitious environment for seeds and acorns to unbolt, take hold in the soft earth, and grow into a forest. All memories of the pond will fade and be replaced by this slow advance of ecological succession.

In September of that year, I set up a camouflage tent blind to watch the egrets dance while spearing fry. The wading birds hunt the shallow end of the pond, but they like one place more than others—a slightly deeper wedge of water and dark mud at the mouth of a bay once three feet deep. Here showers of fright tear the surface in response to a bird's ardor. The fry, who a few weeks earlier had been chased by knuckles of water created by bass now too large to pursue them, use the contours of the bottom, the depressions, the elevations, the minute furrows in the mud as alleyways that lead to higher or lower elevations, a rationale dictated by self-preservation.

I sit on a swivel plastic dove seat and peer at the water through long, narrow openings in the tent fabric. Schools of needle-size fish go on about the business of feeding, independent of each other, until something frightens them. Swimming next to me, they disturb the surface like hair brushed across water. My blind overlooks a nervous habitat.

Every few seconds a silver-colored minnow banks and blinks bright on top of the mud. A kingfisher drops from the sky, its chortle unheard by the fish below, its fall unseen by me until the bird strikes water and creates a circular moment of exultation and terror. The first perfect circle ever drawn was undoubtedly sketched on water.

In the shadows of the woods surrounding the pond, the great white egrets chase each other, wings cupped, tumbling upside down and back up again in a facsimile flight of courtship. They then return to the task of spearing the mud.

With one leg raised and its neck extended, the closest egret pauses, high steps, and then runs to the fry, wings cupped, like a mockingbird chasing grasshoppers. The tall white bird then cocks its head parallel to the water and throws its beak with the sideways motion of a child skipping stones. Without pausing, the egret raises its bill to the sky, a tiny fish struggling at the tip.

It is too small to mollify the bird's metabolism, but thousands more minnows await.

After some time in the blind, a familiar feeling comes over me, a feeling I have felt in other blinds in other states and countries while waiting for geese or ducks to acknowledge my decoys, a feeling that if I leave the blind, something I don't want to miss, something special, something unique, will happen. It's a concern that has always kept me pinned to my seat.

A few hours later yellow jackets fly in and out of the tent. The dog snores. My ass hurts.

TARPON

These days, when I drive south on the parkway to Miami, my car hugs water hyacinth–bordered canals I used to fish fifty years ago. The waterways and land to the east are now shoelaced by the narrow roads and cart paths of golf communities, and the landscape to the west is dominated by massive cement pillars supporting the cables that deliver electricity to South Florida from the nuclear plant built on Hutchinson Island in St. Lucie County. The ponds on the island used to be filled with baby tarpon—small silver-colored fish that took inch-long white streamer flies cast off the tip of a six-weight rod—who over time had acclimated to the landlocked, brackish water. In those days I fished for them wearing beat-up sneakers and shorts; I waded the black water looking for a shallow roll, a fin, a push, any hint of motion. It didn't matter that the antics of the juvenile fish invariably ended inside a tangle of green mangroves.

Back then Florida rewarded those anglers possessing a modicum of resolve with spectacular fishing. A three-hour drive from West Palm Beach north to the Indian River for a morning of gator trout fishing or a five-hour run south to Big Pine Key to look for tarpon was not uncommon. Sixty miles west of West Palm Beach, Lake Okeechobee offered first-rate duck hunting, bass fishing, and the most prolific crappie population in the state. The multitude of inlets on the east coast of

Florida attracted a fall migration of mullet that, due to their numbers, ignited the appetites of tarpon, snook, sharks, barracudas, jacks, and every other predator a mullet could entice. As a bonus, if a school of mullet wandered offshore into the Gulf Stream, they were greeted by marauding sailfish and marlin, king mackerel, and the rest of the pelagic community.

West of Fort Lauderdale, the Tamiami Trail provided access to a hundred miles of waterways filled with garfish and bass, bream, and landlocked tarpon. Even farther south, the Florida Bay bred every fresh- and saltwater game fish a shallow-water angler could dream of. The inshore and offshore waters of the state were some of the finest in the world.

In the late 1960s the fishing grounds of the Keys were the logical destination for someone who lived in Florida and loved to fish. Key West had not graduated into the T-shirt and cruise ship factory it is now, and there was sport to be had, both on- and offshore, with little competition on either hunting ground. I first went to the Keys in 1967 as a client of Stu Apte, a larger-than-life figure during those early years of fly fishing. I had met Stu and his wife, Bernice, at Deep Water Cay. In March of the following year, I made my first of hundreds of oversea highway trips to the Florida Keys, a destination that altered the course of my life.

Having fished the Bahamian waters for a decade, I understood how the light, winds, and tides altered the mosaic patterns of rock and grass on the floor of a flat and the feeding of its fish. I had jumped summer tarpon traveling in and out of blue holes and soaked mullet for them in the passes and inlets of South Florida. But I wasn't prepared for what Stu introduced me to in the Keys. Mile after mile of shallow, pastel-colored water and blue-green channels interrupted by dozens of keys, green and low to the water, overflowing during the nesting season with frigate birds and pelicans. And, while the winds

could be taxing and the cloud cover relentless, by the middle
of March the temperature had warmed the Gulf waters suf-
ficiently to incite schools of tarpon to migrate north into shallow waters, driven by the strength of the tide and the phase of
the moon.

Tarpon (*Megalops atlanticus*) look like an oversize version
of the threadfin herring I now release in my pond. While they
are related to eels and ladyfish, in time, and under the right
circumstances, tarpon can grow to eight feet long. Flat bony
plates, as exaggerated as the carapace of an armadillo, shape
their prehistoric-looking heads. Overlapping blue-gray cycloid
scales up to three inches in diameter coat their figure. The lower
jaw of a tarpon juts resolutely beyond the gape of its mouth,
bestowing upon the fish a truculent look that is mercifully off-
set by large, anxious-looking eyes. Underwater the fish shim-
mers like a huge hoary ghost; in the sky the fish flies above a
column of shattered water as unbridled as a bird, the strength
of the sun affirming its strength and beauty.

Discounting those fish sighted in Nova Scotia or off the
French coast of Brittany, tarpon live in the warm, temperate
climates of the Caribbean and Gulf of Mexico. A small percentage of summer tarpon migrate as far north as Virginia and as
far south as Brazil. Across the ocean they are found in Africa,
from Senegal to South Angola, all destinations bleached by the
sun.

Tarpon that live in the inland bays and southern marshes
often take on a brown or brassy color. On the flats they show
up dark blue to greenish black. In low light they transform
themselves into shadows. When hunting schools of mullet or
menhaden, perhaps even to escape predators, and certainly to
protest a hook, tarpon use their scimitar-shaped tails to propel
themselves into the sky. Furthermore, tarpon almost certainly
jump because jumping is fun. Their ability to fly, added to their

size and strength, raises them in the minds of most fly fisher-men to the premier trophy in the sea.

The tarpon we know today swam in the ocean eighteen mil-lion years ago (humans did not branch away from chimpanzees until six million years ago.) The fish is a modern-day fossil that wears eyelids and breathes through its mouth. Their lives span decades, and the females lay up to twelve million eggs a year. Painted by Michelangelo, a tarpon appears to be featured in the Sistine Chapel.

I love watching tarpons eat flies. That a hundred-pound fish would turn itself inside out to swallow three inches of feathers is a wonder. Since many tarpon weigh as much as a runway model, their interest in tiny morsels of food should not surprise me. Still, I am amused when a tarpon takes a fly, because the taking represents an error in judgment one assumes the fish is going to correct but doesn't.

The approach a tarpon takes to a fly is particular to the indi-vidual fish and depends on the visibility of the water, the action of the fly, and the competition. I have watched flies disappear into the mouths of tarpon unlocked to the size of manholes, and I have seen flies tenderly sipped through hesitant lips. Some fish hit the fly with the resolve of a left hook, while others lunge at the offering in a manner that ignites the sea, the sky, and, if he exists, God himself. Hurling itself out of the water, the fish sails through the air in a paroxysm of head shakes intended to dislodge the intrusion in its mouth. The violence of these out-bursts turns the imagination of the angler inside out.

A five-year-old tarpon migrating up the Keys weighs fifty pounds, give or take. The largest tarpon caught on fly was roughly fifty years old and weighed just over two hundred pounds. Somewhere between those numbers are the size and age of the tarpon I fished for in 1967—big, powerful fish that matured slowly and, just like a man, put up a variety of fights.

I once released a 120-pound tarpon, burned out at the boat, in fifteen minutes—a fish lacking stamina—but I won't forget struggling with another fish, a much smaller—90-pound—tarpon that did not jump but swam doggedly to deeper water by worming a path through the spans of the Seven Mile Bridge for two hours. I recall the determination of the fish, which perhaps had sensed my intentions. It was a sinister quest, as I had intended to kill that fish for the sake of taking a picture. I'd planned to hold the bleeding tarpon off the bow and pull it partway into the skiff the moment a tiger shark or hammerhead came after it. I knew that the shark would follow the tarpon out of the water and that the picture would be unique. Once gaffed, the tarpon was set upon by three skulking bull sharks before the camera could be readied. I have not forgiven myself for that bit of hubris.

I fished with Stu Apte for some weeks before moving on to a number of other guides, some good, others less so, some looking for four o'clock, and at least one who became a close friend. Stu was an excellent guide with a deep understanding of tarpon, their habits and preferences. He also had a propensity to shout instructions at his guests. Having spent eleven years in boarding school, I was weary of amplified commands and hated being yelled at. To his credit, Stu doled out his invectives in an effort to teach his clients how to fish, and his yelling was therefore not gratuitous. Unfortunately he would lean on the words and keep leaning on them until he was apoplectic. The rest of the time, in the boat and onshore, he was charming. Fifty years ago, just like Jimmie Albright and others, Stu was considered "difficult."

My predicament was snatching the fly away from tarpon before they had a chance to close their mouths on it. That year, I remember only catching four tarpon with Stu, none over seventy-five pounds.

One afternoon, on the face of Big Spanish Key, after missing three fish in a row, I encouraged Stu to take the next shot and handed him the rod. He made the right cast and fought a big, beautiful female, which after only twenty minutes I gaffed. Hanging from the Big Pine Key scales, it weighed 152 pounds, breaking Joe Brooks's record by 2 pounds. I had and still have no regrets about handing Stu the rod. If by some miracle I had hooked, fought, and caught that fish, the result would have been a farce.

It has always been a source of wonder to me how rich men, powerful, dictatorial men—in most cases assholes—allow themselves to be treated like shit by their guides. I have watched CEOs hang their heads and blubber like sows in response to the verbal abuse from guides who, for the most part, have no idea what a spreadsheet looks like. Such men spend their entire vacation being brayed at for not placing a bunch of feathers, a shrimp, or some hardware where they were told to. And, at the end of the day, they invite their tormentor to dinner.

In another act in this theater of the absurd, in many of the lodges across the Bahamas that I have visited since fly fishing has become fashionable, I have heard these captains of industry, after a few drinks, brutishly insult the young executives they had invited on the company's dime to "fish with the boss" in front of the rest of the guests. The next day they, in turn, take the same abuse from, say, a Bahamian guide who can't spell his name. There must be a medical term for such behavior.

What I took away from my first tarpon season with Stu was a better understanding of the migration of those pelagic fish in specific depths of water and the tides that impelled them. I also learned how to fight a big fish on a fly rod. A master at getting tarpon to the boat quickly and efficiently, Stu knew how much pressure he could ask of his tippets, because he had tested them by pulling weights off the floor of his garage. He defeated tarpon psychologically by never allowing them the thought of freedom. Stu applied maximum and constant pressure and physically turned fish to where he wanted them to be, against their will. A boxer and fighter jet pilot, he understood the weaknesses of men and beasts.

While in the Keys that first season, I befriended Stu's partner, Woody Sexton, a product of the California school of steelhead anglers, who drove east and south to the Keys every winter to guide for tarpon. We quickly became friends, but when I asked him to guide me the following year, Woody followed his principles and declined. He felt it would look like he had stolen me behind Stu's back. I booked as many guides as I could for the next season and fished with Woody in 1969 and for many years thereafter.

Woody knew from Stu of my tendency to pull flies out of tarpons' mouths, and he calmly explained to me that I was too quick on the strike, that I should take my time. It was, he said, like shooting too early into a covey rise. Thankfully by the time he and I fished together, my problem had resolved itself. By then I had seen tarpon follow flies, sip flies, rush flies, grab flies, gobble flies, scarf flies, refuse flies, and spook from flies, and I had learned to be patient.

In the course of fishing with a multitude of guides during my first few seasons in the Keys, I became familiar with the marine landscape from Marathon to the Marquesas, specifically the variance of tides and the effects of more or less water on the habits of fish. When those feelings became instinctive, I bought a boat, a Maverick made by Wally Cole in Miami. Originally built as a racing boat, the seventeen-foot Maverick had been transformed into a flats skiff. Although by today's standards the skiff would be considered a beast to pole—she weighed 750 pounds—back then she was the pearl of the fleet. With clean lines and a racing hull, she took waves like a surfer.

However, like the lady she was, she did not respond well to ham-fisted orders. The lady and I made hundreds of trips across the deep water of the Boca Grande Channel to the Marquesas, and a couple of trips from West Palm Beach across the Gulf Stream to the Bahamas, a fast two-hour summer run of sixty miles, which was always an adventure of its own.

I had ripened into what guides referred to in a dismissive tone (and, I would like to think, with some envy) as a "do-it-yourselfer," someone who could afford his own skiff and the gas to make it run. The few people like me who fished the lower Keys in the early 1970s were not popular within the guiding community, small as it was. In their minds we were taking fish off their table, crowding the habitat, being a bother. In truth, Key West boasted only two full-time flats guides, and we who fished for pleasure were blessed with tens of thousands of acres of spellbinding shallow water and a flats fishery for bonefish, permit, and tarpon that was unsurpassed in the continental United States.

In 1973, a few years after I started fishing with Woody, I helped my brother-in-law and filmmaker, Christian Odasso, produce a documentary on tarpon fishing. The film was shot in 16mm film in Key West and on the surrounding flats. One of the cameras we used drove the film at four hundred frames per second. Against the transparent backdrop of the sky, every gill rattle, leap, spin, and other contortion the tarpon put its body through to shake loose the irritation in its mouth was slowed down fifteenfold. Protracted in time, the visual distortions were exquisite, just as gorgeous as it would be if one were to slow down the dips and pirouettes of a ballerina. At speed, the tarpon's leap into space is always a magnificent display of aquatic lunacy; in slow motion the gyrations resemble the slow, enfolding movements of a dancer practicing in front of a mirror.

For almost two decades I invested hundreds of hours poling the shallow flats below Key West looking for disturbances,
troubled water, the roll of a bronze head, shadows, apparitions,
strings of motion, the glint of an incandescent gill plate. The
tarpon migration was motivated by the change of seasons and
the arrival of the warm tides that filled and emptied the flats,
creating shifting depressions and passageways for the fish to
use. Once upon a time, I knew the tarpon landscape west of
Key West as well as I now know the landscape of my pond.

WOODY

Woody Sexton and I sat next to each other in the stern of his wooden sixteen-foot Nova Scotia skiff waiting for the cloud that shaded the morning sun to pass. A soft gray mantle rested on the flat and there seemed little point in poling, so we waited—engine idling, the bow troubling the surface of the water—for the sun to escape and expose the flats surrounding Coupon Bight.

It was June, and we had been fishing for fifty-five days in a row. The skiff felt tight as a glove. The daily repetitions of launching and running, of poling and looking had elevated the smallest, most idiotic hint of humor to the level of high comedy. I followed the last leg of a mosquito's journey from the mangrove shoreline of Big Pine Key to the skiff and watched it sink its stylet into Woody's neck.

"Goddamn bloodsucking son of a bitch," he said, jumping up. The skiff tipped to starboard. He took a shot with his hat, missed, and rambled on about malaria, dengue, yellow fever, encephalitis, and other insect-related diseases. Woody used words like *C. pipiens quinquefasciatus* for a house mosquito and *Lymphogranuloma inguinale* instead of VD. His forearms were knotted like a sailor's; his face, a reasonable facsimile of the western landscapes he hunted each fall.

Endowed with a critical mind and a near-perfect memory, Woody Sexton graduated in 1943 with a degree from Humboldt

State University. Diploma in hand, he went into the US Navy, and though he did not see action in World War II, he did witness the nature of his species at close quarters. Woody might have been a doctor, a lawyer, or an engineer, but instead, after being released from his duties, he made his way from San Diego to the redwood forests of the Northwest and became a lumberman. For a decade he lived in a world defined by fulcrums and angles, exposed to the raw, often dangerous physicality of his profession. Far from the confines of civilization, he burgeoned into a "master lumberman," skilled in the art of felling trees bearing the circumferences of school buses to an exact place in space, a designated piece of ground where other—less skilled— men could get to the wood quickly and dress it for sale. In his off-hours Woody read books.

We had both celebrated birthdays in May. I had turned twenty-five, Woody, forty-seven, a fact that did not please him; I was too young to care. To retain a semblance of the physical strength he had developed during his years in the woods, Woody kept in shape by completing two hundred push-ups and two hundred sit-ups every morning. He would then twirl a forty-five-pound cement block tied to a rope around his body like a hammer thrower for fifteen minutes before we met for breakfast. One afternoon, after pushing his skiff for six hours, I watched him squat behind an unsuspecting guide who weighed upward of three hundred pounds. Woody wrapped his arms around the man's knees and picked him straight up off the ground. He grinned like Popeye the sailor.

When the tides were wrong or the fish elsewhere, we talked about everything from river systems and ocean currents to books by Roderick Haig-Brown, migrations, bird hunting, and gear of all kinds, from bait-casting reels to Woody's navy-gray skiff. Quick and light over the water, she was a dream to pole, which more than made up for the fact that because of her round chine she was wet and tippy. The skiff had been built in Islamorada, Florida, from a cold-molded wood hull manufactured by the Chestnut Canoe Company in Nova Scotia. From inside her confines we had jumped 150 tarpon since the beginning of the fishing season in April.

Outside the skiff the water was syrupy and dark. Light fanned toward us in harmony with the cloud struggling from under the sun. The mosquito headed for shore. My guide, whose friendship I grew to cherish, shoved the fishing cap back on his short white hair and sat down. All at once sunlight spilled over the flat and Woody relaxed his grip on the tiller. Hot humid air enveloped us—a rallying call for tarpon to migrate from the comfort of deep, offshore water to the shallow flats that buttress the southern United States.

Woody allowed the skiff to lose momentum. We took turns sitting and standing, both of us watching from different angles for disturbances on the surface, for shapes inside shadows, for distortions in the slow, oily rhythm of the tide, for motion of any kind; we watched for rings, for daisytails, for the surge of water that shepherds the broad heads of shallow-swimming pelagic fish.

Woody raised the 40-horsepower engine manually, unleashed the push pole from under the bungee cords that bound it to the chocks, and moved to the bow. He maneuvered us into the sandier, shallower water of the northeast end of the bight, where on full-moon tides schools of smaller tarpon painted coin-colored shadows on its white sandy floor. The big fish favored the deeper water to the southwest, where they slept like fat old men inches above the turtle grass. I pulled fly line off the reel. The skiff rose and fell with each stroke of the pole.

Woody poled with the same focus he devoted to bird hunting. "I miss the smell of mountains," he had said two weeks earlier, meaning that his season in Florida was coming to a close. He was now dreaming of anadromous fish and mountain-climbing birds. In a week, maybe two at the most, he'd fill the cab of the Ford pickup truck with his Spartan possessions and drive to Northern California. There he would look for housing in a small, out-of-the-way town with a good library. In the fall he'd drive to Idaho and fish for steelhead on the Clearwater River. Later, when the aspen brightened the foothills and the mountaintops displayed the first signs of snow, he would carry a shotgun high under the clouds and chase elegantly feathered birds imported long ago from Europe and Asia.

Woody worked as a flats guide in the Florida Keys for five months a year in order to spend the following seven in the wilds.

"I watched the entire side of a mountain run uphill," he said one day as he poled the face of Loggerhead Key. He was referring to an October morning some years back when he pushed two thousand chukar partridge into flight off a mountaintop south of Orofino, Idaho. Another time he described a foggy morning on the Bryan Pool of the Eel River and of landing five steelhead on a Polar shrimp fly, each fish weighing between twelve and eighteen pounds: bright, powerful sea-run rainbows, benevolent hosts to the sea lice they ferried to shore.

Woody came to the Keys in 1959 from those rivers of California—the Mad, the Eel, the Trinity, the Klamath, and the Smith—where he and other young men of his generation braved cold, chest-deep water for the chance to hook silver-colored fish fresh out of the ocean. Men who hurled three-hundred-grain lead-core shooting heads into deep, heavy water from dawn to dusk. Men who were seen by other anglers as monoliths, fixtures in the rivers they fished.

Warm weather, shallow water, and the reports of tarpon by the thousands drew Woody and his peers south, much as gold had drawn their forefathers west. Once these river anglers adapted to the wind and converging vectors of fish and boat, the mechanics of casting a fly while standing on top of the water—as opposed to in it—became second nature.

"In those days, during a spring tide, we expected to see two, three, maybe four hundred fish in Coupon Bight," Woody said. "Now, with the boat traffic and the commercial real estate development, we're lucky to see thirty."

He cupped the palm of his right hand around his sunglasses and scanned the water from under the bill of his hat. "That first year my partner, Jim Adams, and I used Fisher blanks and Jimmy Green fly rods," he added. "Our reels were made by Young and Hardy. At night we hung from bridges and floated flies on tides that flowed like rivers. A twelve-foot aluminum boat powered by a 7-horsepower engine drove us to the flats. We poled with a borrowed curtain rod. All we knew about tarpon was that they ate flies. We applied ourselves to that premise and cast at their faces. It made for good adventure."

As someone who had spent the better part of his life on the edges of civilization, Woody took issue with those he viewed as degrading the natural world. "Politicians and lawyers are ticks," he'd finalize.

My sporting life before meeting him had been European in attitude, an ethos that translates loosely into: "Kill everything that moves." My new fishing friend and mentor was more interested in the health of his surroundings than in the ambitions of men. In time he taught me to respect the playgrounds we played in.

Since his first trip to the Florida Keys in 1959, Woody met all the great and not-so-great anglers, boatbuilders, and rod-and-reel makers of the era: Apte, Brooks, McNally, Lee Wolf, George Hommel, and Captain Mac, and dozens more befriended Woody, whose acumen was more developed than theirs. Later,

when a younger generation of men such as Gil Drake and Steve
Huff moved to the Keys to fish, they too delighted in his company. Woody died in 1998 after a long and disturbing bout of
dementia, not a disease I ever thought would afflict Woody Sexton, whose mind had always been as sharp as a mind can be.
Huff, who carries the mantle of best "pound for pound" guide
ever, scattered Woody's ashes over the surface of Loggerhead
Key—a good and fitting resting place for a special man.

"Fish!" Woody pointed to a disturbance one hundred yards
down light. He wedged the pole against his hip and spun the
stern of the skiff toward the commotion. In the distance the
water chortled as if pressed over boulders. Moments later early
light fell on the chiseled heads of fish the color of bronze.

"Ten, maybe fifteen tarpon. Coming at us!"

The orange-and-yellow fly, tied to be fished in dark water,
was wet in my hand, slick as a worm. One by one the tarpon
surfaced—some with purpose, others playful—all causing the
sea to part, each roll imposing levels of apprehension to the
rest of the school. I rocked my right wrist back and forth, back
and forth, forcing the belly of the line to roll and pull against
the fly.

The closer the fish swam into range, the more the notion of
time faded. I heard the rub of Woody's hands against the push
pole and felt the skiff yaw to the right, inviting my backcast. The
fly landed a leader's length ahead of the lead fish. A mouth rose
cavernous out of the water and closed. I raised the rod, ran out
of striking room, and watched the fly sail gracefully through
the air. It fell in a melancholic tangle on the surface of the water
next to the engine.

"Shit! Shit and shit," I yelled as I watched huge puckers of
creamy water rise from the floor of the flat.

"Shit" I said again. Woody smiled and said nothing.

I poled and Woody fished. His regular fee was sixty dollars a day. He charged me thirty, we shared the fishing, and I brought the lunch. That was our deal.

Some days we trailered the skiff and chased bonefish and permit below Key West, but Big Pine Key, Summerland Key, Cudjoe Key, the Seven Mile Bridge (during the palolo worm hatch), Monster Point, and the Eccentrics, where tarpon would swim out of the deep channels, looking like thick black eels, were the settings Woody was most comfortable fishing.

Every morning, with a notion of sunlight angling off the blacktop into the windshield, I would drive from Key West to Summerland Key and meet him at the Chat and Chew Restaurant. After breakfast, which often consisted of a foot-long hot dog, I'd follow him up to Big Pine Key, where he kept his skiff.

In the afternoons the drive was tougher, the sun brighter, my eyes tired from probing through layers of seawater, overloaded from the incongruity of watching huge fish fly across the sky.

Willie Mae, my wife's housekeeper, would greet me with a drink on the doorstep of the conch house on Duval Street we rented each spring. The children ran up from the street and hugged my legs. Willie Mae was a large and reassuring woman with shiny black skin. She wore a permanent smile and had a gift for making memorable chicken sandwiches. If I wanted sympathy regarding the hardships of my sporting life, it was to her I would turn.

Woody had the high casting motion of a deep-river wader. On the backcast, his right hand hugged his ear before rising straight up to the sky. His double haul was concise and powerful. He was at his best when he saw fish late and made the throw purely from instinct; his forte, the short cast. A long string of rolling tarpon noodling up to the boat from a distance created an intellectual conundrum and therefore a pause, much like the pause of a shooter questioning how far to lead a duck crossing the backdrop of a bluebird sky. Woody liked to snap his casts just like I liked to shoot a gun—quickly.

I poled the north side of Coupon Bight, the sun warm on my shoulders. The turtle grass seven feet below the hull chased the tide. Shallow coral heads scattered under the bow, adding to the difficulty of poling. For the sake of silence—the clang of a pole's hardwood foot hitting rock shatters the nerves of large, sleeping fish—I could not drive the pole with the speed I would have liked. Instead I had to catch it before it reached the bottom and ease it the rest of the way into the grass. My hands hydroplaned on a permanent film of water. The power strokes ended at my knees.

Two tarpon swam parallel to the boat, a foot under the surface. Woody made a quick half-moon cast behind his right shoulder and drove the rod toward the fish. The fly turned over a few feet ahead of the closest tarpon, a fat eighty-pound female, followed by a smaller, leaner male. Both fish rose to the fly in a confusion of bodies. A large, round scale spun luminous to the bottom. The smaller of the two fish knifed cleanly out of the water close to the boat.

"I never struck him," Woody exclaimed. "Lots of excitement for a two-inch meal!"

A third fish, deep and more difficult to see, inched toward us on the bottom of the bight, sixty feet from the skiff. Woody retrieved the fly and made the cast. This time, when he stripped the fly, something slow and confident rose from the grass, settled behind the fly, opened its mouth, and ate it.

The size of the fish in relation to the fly and the boat didn't fit the proportions of the tarpon we were accustomed to; the largeness of its mass and purpose of motion was unsettling. The fish turned from the boat, and Woody's white Shakespeare fly rod bowed. Instants later the ocean came alive as a tarpon as broad as the wing of a plane cartwheeled through the surface and into the sky—a silver fish with golden eyes rising, shedding light, bending the scales of our imagination. At the apex of its jump, pinned by gravity, the tarpon seemed to hesitate for an instant before tumbling like an anvil back to the sea. Yards of white water leapt out of the water.

The fly rod in Woody's hand looked small, ludicrous. He turned, his eyes round under his sunglasses. He asked: "What do you think, two hundred pounds?" His lips were stretched tight against his teeth. I dropped the engine and pulled the ripcord.

"At least!"

I motored after the tarpon, guided by the tip of Woody's rod. Again and again this dancer of a fish sailed through the warm humid air, dripping with light and power. Except for a handful of blue-and-black marlin, some sharks, a whale once, I had never seen a fish of that size before.

And so it went: Eleven times the tarpon hurtled through the glasslike surface of Coupon Bight, eleven jumps into a foreign medium, eleven back-breaking falls. The fish dragged us to the entrance of the bight and stopped. Woody reeled. Sweat ran down the middle of his back. To keep the slack out of the line, I worked at outguessing the fish's moves. Woody reeled the backing back onto the reel. The angle of the fly line steepened. Beads of water fell in long thin strands into the bight. The giant rolled and lunged partially out of water one final time. It shook its gill plates in exhausted fashion. When it fell, it fell slowly, and then laid on its side like a timbered log, benumbed, the tips of its fins quivering.

We instinctively knew that our only chance to land the tarpon was to get a gaff into it quickly, that if the fish regained its senses the twelve-pound test leader would lose the ensuing fight. I idled the skiff as close as I dared and, reluctant to disturb the fish's stupor, cut the engine. Woody reeled the tarpon closer to the skiff. It looked twelve feet long. I reached for the gaff we kept tethered to the gunnels. It wasn't there.

"Woody! We left the damn thing at the dock!"

We had both killed tarpon before, mostly for other anglers, and hadn't liked it. We had agreed a few weeks earlier never to kill another one. A tarpon is too much like a St. Bernard, a generous fish with a huge heart and a gift for jumping, a fish with a past, a fish with a soul. Who, except for a fool, would deliberately kill a soul?

But the truth was that neither of us had ever imagined that a trophy, fifty pounds heavier than the world record, would one day rest within arm's reach of our boat. I rummaged in the bow and found the lip gaff.

"When I get it in its mouth, we'll pull him in," I said, reaching over the side. The tarpon dropped out of sight, as did the tip of Woody's rod. Slowly, carefully, he raised it out of the water. The fish moved. The odds were against us. Even with a gaff hold in its mouth, the tarpon would turn the skiff over when we tried to pull it in. However, we would try, not for the killing, maybe not even for the glory, but because it was an integral part of the game we were playing.

I leaned over, my right arm completely underwater. The gaff glanced off the bony side of the tarpon's head. A broad, dark-rimmed tail swiped at my arm and kicked the fish back to the bottom. Woody cajoled it slowly back to the surface. This time, though, the tarpon—perhaps waking from a dream—sensed an unholy presence and resisted. Woody applied a little more pressure and the rod weathervaned in his hands.

The fish was off.

The big tarpon canted this way and that on its way to the bottom, righting itself just above the turtle grass. We watched its broad dark back melt into the shadow and dreamily, majestically swim out of our lives. On one hand, I despaired at how close my friend had come to angling immortality. On the other, I felt an odd sense of relief.

Twenty years later, Woody told me that he still dreamed about the fish that had somersaulted across Coupon Bight, that he could still feel its power and majesty in his hands. I now know that the feeling of relief I felt that morning was motivated by envy.

There was nothing to say and nothing left of the day. We sat in silence, reliving the moment that had shaped the longest and shortest ten minutes of our fishing lives.

The wings of a cormorant ticked the surface of the water. A Navy jet howled across the sky, and the wake of a distant steamer passed under the hull of Woody's skiff. In the distance I could hear the manic voice of a flats guide berating his client.

A school of tarpon rolled into range. We did not stand up.

FALL POND

In October the heat of summer goes to ground. Gone are the blooms of algae that surfaced green and ethereal to the surface of the pond in July. Everything is crisp and sharp now. Early light delights in chasing well-defined shadows out of the woods. The view from my chair offers a glimpse into the landscape's final act of decadent splendor in the pause between the green abundance of summer and the fossilized colors of winter. Wild sunflowers, partridge peas, black-eyed Susans, and milkweed border the pond in a riot of overly ripe blossoms. Cattails lose their color to the season and their seeds to the wind. The cypress trees flaunt themselves in the shallows of the pond in a final display of pageantry. Beyond, goldenrods take the stage and initiate a progression of yellow wildflowers that persist in invading the fields. Crotalaria, banana water lilies, foxglove, tick seed, and marigold infuse the sedge grasses up to where clusters of more partridge peas and rows of planted lespedeza entertain the quail.

The light caresses the surface of the pond with golden engravings early and late each day. The temperature at noon has dropped ten degrees in the past week, and my gait as well of that of my dogs has assumed purpose. The pine trees cast narrow beams of shade deep into the pond. A tumble of acorns belonging to the live oak tree next to the pond house carpets the ground next to the dock; some find refuge in my boat. Gray fox squirrels feast on the yolks wrapped inside the kernels.

An east breeze maddens the water, swirling and snarling and setting off pond-size windstorms. Hundreds of shiners replace the bream under the dock. The bass rush at them with disheartening results.

Hummingbirds, the feathered bumblebees of summer, having abandoned the sugar water I extended to them during their stay in North Florida, are on their way to Mexico. Fifty miles south of me, tens of thousands of Monarch butterflies swirl around the St. Mark's lighthouse, in advance of the east wind that will set in motion their annual flight across the Gulf of Mexico. The red-winged blackbirds have been replaced by ruddy ducks that bob on the surface like brown Ping-Pong balls. Squadrons of teenage bass chase the fingerling shad and disrupt the countenance of the pond.

Fall fishing in northern Florida combines the best of spring and summer fishing—hot afternoons translate into surface action, cool mornings encourage swimming plastic lures deep. At times, when the activity is heavy and the frightened shad are scattered over the surface, I'll cast a hookless rubber surface lure in the mix and count how many strikes I get before the lure is back at the rod tip. The most bass I've had hit the plug in one cast is thirteen.

More than the fishing—or in this case the teasing—what sends me back to the pond and causes me to linger on the water at this time of year is the reflections of the cypress trees on its surface, the medley of ginger-colored branches layered like the dress of a flamenco dancer, the golden colors that careless shivers of breezes confuse and greater winds render abstract. At four o'clock in the afternoon the fan of fall colors that rises from the floor of the woods surrounding the pond is brightly reproduced on the water. Expecting a percentage of loss to disease, beavers, and the occasional drought, Bill and I planted five hundred seedling cypress trees before the first big storm that filled the pond. Twenty-five years later, one out of every eight trees sown casts its brilliance into the pond.

The pond's autumn glory is my idea of splendor.

A month later a mustache of dead, russet-colored needles appears tucked in the branches of the loblolly pines, interrupting the trees' concord of green. Each morning I interrupt the same fox squirrel sitting on its haunches enjoying his kernels next to the transom of the boat. White egrets push and prod the mud for food, their necks extended in anticipation. A cold breath of northern air momentarily excites them into a facsimile dance of courtship. Yellow jackets rise and fall, uneasy with premonition. The wildflowers wilt.

As the light weakens and the temperatures fall, I see the pond's residents gorge themselves in preparation for the physical demands of migration or the abstemious mandates of hibernation. Overwintering at this latitude is not as demanding as it is farther north. Nevertheless, in preparation for a few weeks of metabolic bliss, snakes and alligators seek specific hideaways in which to indulge in the slumber that will follow the frenetic activities of summer. At this time of year, Olga retreats to the comfort of mud at night. When the temperatures drop into the forties, I won't see her. But on sunny days, after a warm-up, her head reappears briefly, looking grim. After a three-month carnival, the pond remains a landscape in motion, though now it dances to a slower beat.

Every year in late November, a day comes when fingers of silver light race across the pond ahead of a cold front. The ensuing rain and wind drag the last of the wildflowers to the ground, and seemingly overnight the clay earth rises, visible through the dying grass. Without waiting for the solstice, winter makes a brief three-day visit. But when the wind replaces the rain and the sky returns to its previous uniformity, fall returns to the pond.

A few late-blooming crotalaria flowers coexist on the stalks beside black pods filled with next year's seeds. Some days the chill in the air invites the shotguns to be unsheathed, the dogs to wear bells, and the quail to be alarmed. A small green water snake cuts across the surface of the pond. I hope for a meeting of snake and bass or snake and Olga, but the transit is uneventful.

Following an increasingly cold afternoon, rain appears from the west and soaks the exposed half of the loblolly pine trees. Olga disappears. The storm passes. Spanish moss drips from the branches of the oak trees. The next day the silky heads of plume grass reflect a clean, bright light. Gray-brown ruddy ducks follow the bouquets of each other's white tails. Reflections on the surface of the pond are pared down to pine trees and sky. I look ahead to the arrival of the ospreys, a silly thing to do.

At my age the last thing I need is to hasten time.

CURTIS

The pond's record, a bass weighing 14.25 pounds, was caught in May a decade ago by Curtis Henderson on a live shiner. A forty-year-old former fishing guide with an aptitude for cutting trees, Curtis had just started working for me and was not aware that I didn't allow live baiting, the most efficient method to hook mature bass and also the quickest way to deplete a pond of big fish. Had the hen bass been caught before the spawn, she would have pushed 16 pounds.

The sun had set when he opened the lid to the live bait well in the stern of his boat.

"It's the biggest bass I ever caught," he said, beside himself with excitement and wanting badly to keep it.

Bill Poppell and I peered into a dark watery hole and at first saw nothing. Then the hole moved. We all guessed three or four pounds more than the actual weight it turned out to be. However, it was by far the biggest bass I had ever seen. Curtis mounted the fish, and it now hangs in his house alongside two hundred or so other mounts.

"What I do is fish and hunt," Curtis replies if questioned.

A formidable understatement, if one accepts fishing and hunting as what people do out of a weekend boat or in the field alongside a Labrador retriever. Curtis is not the only man whose approach to field sports reaches back to an ancestral drive to put food on the table, but certainly his proficiency at killing game is at odds with the cultivated sentiment of the modern bourgeoisie.

Curtis is a midsize man with a mane of blond hair, a Gallic nose, and an engaging smile. He doesn't smoke or drink—not even coffee—but he eats 750 doves a year, 100 largemouth bass, half a dozen wild turkeys, and 2 whitetail deer from Texas (where they grow big). Game is what Curtis feeds on, along with canned vegetables and soft drinks. Over the years he has developed what is referred to down here as a Pepsi belly.

Curtis is the best dove shot I know. The birds he aims at die in the sky. He uses a twelve-gauge Remington 870 for all his hunting, and he points out to those who consider this excessive that his intention is to kill, not cripple. Curtis does not wear a hat while hunting and he doesn't move once stationed. It is as though the birds (ducks included) are mesmerized by his blond hair and fly to wherever he's standing.

He once spent a season bass fishing on the professional circuit, but he became impatient with the people, the advertising, the bragging and posturing. So he returned to Lake Jackson, where he grew up. That is where I met him.

A few months after he'd caught his big bass, Curtis reeled in two double-digit fish on successive rubber-worm casts at the mouth of a small bay west of the pond house. The air was hot and humid, as hot and humid as Curtis became when I suggested he release the first fish. It weighed 12.75 pounds.

"I ain't ever put back a fish like that before," he stammered.

"You kept the fourteen-pounder," I reminded him. "You have to catch a bigger one before you can keep it."

Curtis's second bass that afternoon weighed 11.5 pounds, and back into the pond it went.

The fishing protocol we arrived at, after a few years of indiscriminate harvesting, is that those bass caught in the pond that weigh less than two pounds are culled; anything over that is either mounted or released. Young bass taste sweeter out of the skillet than the older fish and, since they are nimbler and more efficient at gathering food than their parents, they rob the pond's matriarchs of feeding opportunities.

When I first met Curtis he lived in a converted tobacco barn inside of which he showcased hundreds of mounted animals he had shot or caught. "They are beautiful," he said. "You like your paintings, I like my mounts." Curtis sees taxidermy as art. Individual mounts remind him of specific moments of his life, just like a painting reminds the painter of a particular time and place.

In both the tobacco barn and now in the new cement-block house he built a few years ago, there is not one square foot of free space. Rows of deer-head mounts line the walls. The horn mounts occupy a room of their own, while the loose antlers that he finds in the field are piled up pell-mell in a corner. Fifty of his biggest bass—including the 14.25 pounder—command a wall of their own. One hundred more bass, all over 8 pounds, hang from the ceiling of his living room. Every deer mount (also one elk, one moose, and a red stag) nudge the Boone and Crockett scale, or at least would make the average nimrod delirious with joy. Ducks, doves, quail, geese, and pheasant fill the open spaces on the walls, while snakes, alligators, wild cats, and foxes live out their resurrection on the floor.

A while back I asked Curtis about women. He was living in the tobacco barn at the time (by nature a dark building). He said that the last woman he dated worked for the Fish and Wildlife Service.

"Off work, she smoked dope. One night when nature called, halfway to the shitter, she bumped up against one of the alligator mounts and freaked out! I'm not kidding you, she completely freaked out, yelled, and peed her pants and everything! I can't put up with that kind of behavior, so I don't date anymore."

I didn't question his preferences in interior decorating.

After graduating with honors from high school, Curtis retired his academic career and began poaching the tenderloin of earth that conceals the blue-blood plantations between Tallahassee and Thomasville. His intentions were extreme. In the cradle of a swamp Curtis would set up for deer and turkey with his 870 and remain perfectly still until it was time to raise the shotgun on an unsuspecting buck or priapic turkey. He was very good at what he did.

Years ago in Montana I knew a Vietnam veteran, a jungle fighter who had returned home angry after two tours of duty. One night four hunters came into the bar where we were drinking. They had had a successful elk hunt, and the four of them drank and bragged and talked tough. After a while the Vietnam vet slipped out of the bar, retrieved a hacksaw out of his truck, and proceeded to cut the antlers off the two elk heads lashed to the hoods of the hunters' cars. Then he returned to the bar and waited. In due course the hunters left, only to return minutes later, screaming their anger. The jungle fighter admitted to shearing the antlers, took them on, and prevailed.

There is not a species on Earth that would want that man—or for that matter, Curtis Henderson—stalking it in the woods with killing on his mind.

OSSO BUCO

Every ten years or so I reintroduce sterile carp to the pond in an effort to prune back on the alligator and coon grass and other invasive aquatic plants that bloom as an ode to nature. Both the weeds and the carp are invasive, but I am able to control the former by introducing a relatively small number of the latter. Creating enough but not too much cover for fry to hide in is better managed by vegetarian fish than by the application of herbicides.

The appetite of a juvenile grass carp is significant; it will eat up to twice its body weight in weeds every day. If the vegetation accommodates the fifty or so I release, the carp will grow from a purchase weight of ten ounces to three pounds in two years, ten pounds in five years. As grass carp grow in stature, they lose their passion for aquatic plants and lose their forage value. Eventually they bottom out the thirty-pound Boga-Grip scale, huge satiated creatures skulking along the shorelines like bankers in three-piece suits.

A bow hunter friend asked if he could shoot a few of the larger, older carp, stretching my credulity when he promised they would be eaten. For a while I watched him in action, dressed in full camouflage as he high-stepped the banks of the pond, acting more shorebird than man. At the end of the day, he offered me a piece of scaled and cleaned carp loin, which

I steamed over lemongrass, coriander leaves, and ginger. The glistening chunk of white meat tasted like lemongrass, coriander leaves, and ginger.

I was watching for fish late one fall afternoon when I noticed a persistent surface activity too far out on the water for me to identify, even with the help of binoculars. It seemed as if something was floating in the middle of the pond and something else was chewing on it. When my curiosity got the better of me, I launched the boat.

The sky was reflected on the water. The pond was dark blue and clean. Small impressions on the surface were being shaped by the breeze. I noticed that what was being eaten was the head of a large grass carp. A softshell turtle, maybe fifteen pounds, and aggressive, was eating the decomposing fish. Strong swimmers, softshell turtles possess long periscope necks, retractable and extendable when the need arises. They are among the ugliest detainees in the pond.

The silver-colored head and gills of the fish weighed about ten pounds, making it by default one of the original 1994 pond releases. The carp might have died of old age or been torn in two by an alligator or even killed by an otter. In any event, its head had been left behind and the turtle was gnawing on it. For some time I had thought about catching a softshell, imagining from locals' reports that it would taste like the green turtles I grew up eating in the Bahamas. When I got back to the cabin I called Skip.

My friend Skip Sheffield, who was raised on a cattle ranch in central Florida, now builds expensive houses in Tallahassee for a living. He wears khaki pants and crisp white shirts during the week as he navigates the condescending attitudes of the prosperous owners, not to mention the capricious nature of their wives, while he builds their dream houses. On weekends, when he is not catching up on book work, he goes fishing.

As a young man, when school was out in Floral City, Skip worked cattle for his parents. At the end of the day and on weekends, he would go squirrel hunting, bass fishing, frogging, and jugging for turtles. "We were poor," Skip told me while he explained how to jug for turtles. "We ate off the land. My dad would give me six .22 caliber shells and expect six head-shot squirrels. I would check my jug after work. If it bobbed I knew I had hooked a mud fish or a softshell turtle. We ate the turtles and gave the mud fish to the black family who lived on the ranch. My mother stewed the softshell meat with rice and vegetables. It tasted like chicken."

As he talked, Skip tied a three-foot-long drop line ending with a 3/0 hook to an empty plastic detergent jug and attached a second line to the jug, tied to a brick. He used a store-bought dough-ball mix to catch the shiner that would serve as bait. Once in hand, he cut the small fish in two and then hooked half of the body to the end of the drop line. Skip tossed the jug, the baited hook, and the brick off the dock where softshell turtles gathered when I fed the bream.

"I'm jug fishing now," Skip said grinning. "Just like thirty years ago."

The next morning I called him. "It moved!" I said. The jug was a long way off the dock.

Skip arrived shortly, accompanied by Huck, his hundred-pound black dog, a mongrel influenced by latent Great Dane genes. Skip, who is sixty years old, is a big handsome man who had recently lost seventeen pounds on a five-week leafy greens rabbit diet. He was proud of himself and insisted he felt great. Nevertheless, since I know he loves to eat and enjoys his wine, I wondered how much longer the diet would last.

Skip laid a gaff in the bass boat, told Huck to jump in, and paddled out toward the jug. Anticipating that he was about to retrieve a softshell turtle, Skip had also placed a plastic barrel in the bow. A setback occurred when he attempted to raise the turtle out of the water. Huck, sensing there was something of interest at the end of the trotline, ran amok in the twelve-foot boat, barking and rearing up on his hind legs. I was hoping that one of them would fall overboard and beget a perfect Laurel-and-Hardy moment, but after a few tippy maneuvers, Skip paddled back to shore with a twelve-pound turtle looking rather like a gravy-colored pancake at the bottom of the plastic drum.

It was not a sight to advance one's appetite.

After a few minutes of obligatory tire kicking, it was time for Skip to prepare his abattoir kill and bleed the animal. An inkling of doubt crossed his face, and he delayed the moment by drilling a screw into one of the wooden braces that held up the dock.

While waiting for Skip, I carried the barrel and the turtle up to the end of the quay and the fish-cleaning station. After wrapping a rag around my hand, I raised the turtle out of the barrel with a length of forty-pound test monofilament attached to the hook. Until that moment I had never seen a softshell turtle up close. The hook was lodged all the way through the turtle's mouth and out the end of its nose. Soon after being suspended, the turtle's neck slowly emerged from its leathery carapace.

"He's tired," Skip said. "He swam the brick halfway to the shoreline."

To myself I thought, *Reptilian brain or not, a hook in the nose has to hurt like hell.*

The head of a softshell turtle is elongated and without character, its mouth is bony, and its proboscis pimply. While swimming underwater it will use its nose as a snorkel in order to breathe. Softshells are carnivorous and aggressive. Everything about them is disconcertingly ugly.

Carrying its own weight by the hook, the turtle grew ever longer as its neck continued to telescope out from its shell.

When it was fully extended, I raised the turtle onto the cleaning deck and held it down. The shell looked hard but wasn't, and it felt peculiar under my hand. Skip, having gathered up his courage, leapt at the animal's extended neck and started hacking. "Got to cut its head off," he said, grimly focused on the job at hand.

More than likely, Skip's childhood memories did not include anything as raw and brutal as sawing off the head of an objecting turtle. Or if they did, the passing years had turned the heroic exploits of a kid into a task older men question.

The neck of a softshell turtle is enormously muscular. Skip wielded his knife for purchase. When he finally managed to cut through the layer of muscle and sinew, he was faced with the problem of locating the space between the animal's vertebrae. His knife slipped and grated over bone. The turtle's legs fought the wooden top of the cleaning station. Skip's face turned red, but with nowhere to go and nothing to say, he applied himself even harder to separating the turtle's head from its body. Softshells don't have teeth; instead they possess hard and sharp upper and lower plates that close like wire cutters. Given the chance, the turtle would eagerly have severed one of Skip's fingers.

All of a sudden the head, like the heads of many of my ancestors during the French Revolution, fell into the barrel, and we were left with a brown, slippery, beheaded turtle. Skip pitched the skull into the pond for the catfish. Neither of us spoke.

He then cut a slit in one of the turtle's rear legs and, holding it at arm's length, walked below the dock and hung it upside down from the screw he had drilled into the brace. It remained there, dangling from one leg for an hour, bleeding onto the grass that had grown under the dock during the time of drought.

When the "Cooter" had rendered its blood, Skip hauled it back up on the cleaning table and positioned it on its belly.

"See here," he said. He pointed at a spot on the side of the turtle's shell behind its front legs. Sure enough Skip's knife slipped quietly into the carapace without resistance. "I enter the body here," he said, "and in the same place on the other side of the shell. I'll cut through the tendons on either side of the body and the shell will come right off."

The cut was about eight inches long and—by far—the cleanest component of the dissection.

"It's been thirty years since I've done this, and I'm not remembering it all." Skip said twenty minutes later as he tried to jerk the soft shell off the body of the turtle. The effort of the entire exercise was visible on his face. "Shit! It's been forty-five years, not thirty," he exclaimed, remembering his age and looking shakier yet.

Finally he separated the shell from the body. Inside were guts and orange eggs and a pulsing heart. But it wasn't over yet. The bones that attached to the turtle's legs would not give way to Skip's knife. Taut muscles and tendons encouraged them to retract and extend each time the blade made a cut. It was a series of chain-saw-massacre moments.

"I don't think we need to kill another one," I said when he finally put the knife down. I wasn't feeling too spiffy myself.

Originally we had planned to kill two more turtles and make a dinner out of them, with wine, silverware, and cherry tarts for dessert. I had been thinking turtle soup, fritters, steak pie, even gumbo. But the grim and protracted process surrounding the *mise à mort* of Lady Turtle caused me to reconsider.

In the end Skip delivered two hind legs, two front legs, and the turtle's neck, having sailed the remainder into the pond. Once he'd completed his job, he left.

In the kitchen that evening, with wine at hand, I boned the meat from the turtle's legs, adding them to its neck already simmering in a broth. The meat of the turtle, now dead four hours,

was still—incredible to me—in motion. A cut here would pro-
voke a reaction there, seemingly without rhyme or reason. The
slicing caused shudders and pulsing contractions that, while
disquieting under the knife, were even more so under the hand.
The turtle meat, about two pounds of it, did not fully relax until
it was laid out in chunks on the cutting board.

I brined the meat in salt and sugar overnight, and the next
day I deep-fried a portion and sautéed a second batch in butter
and garlic. The meat, which quite honestly I looked at long and
hard before putting it into my mouth, was delicious, sweet and
veal-like. The taste carried a slight flavor of almonds.

Because it's what I enjoy doing, I browned and then slow
cooked the remaining meat in a tomato sauce just as one would
prepare osso buco, with mushrooms and pearl onions. When
the meat had softened and the flavor pleased me, I added, off
heat, salt, pepper, and a couple of red Thai peppers. My softshell turtle was accompanied by risotto and gremolata (parsley,
pine nuts, and lemon zest) into which I crushed two more dried
peppers.

Skip went off his diet.

SAILFISH

In my twenties and thirties, when my shotgun was pickled and the dogs were resting from the arduous efforts of a long hunting season, my thoughts turned to pelagic fish that I could tempt using flies made to look like school bait. I liked warm weather, small boats, and fly rods. Penciled right below tarpon on my list of favorite fish were sailfish: the large Indo-Pacific sailfish that journeys north and south off the west coast of our continent from Baja California to Chile, and the small Atlantic subspecies that migrates down the eastern seaboard to the Caribbean every winter.

In the summer of 1968, and again a year later, I joined Gil Drake and his wife, Linda, for a week of fishing in Playa del Coco, a small fishing village located in the northwest corner of Costa Rica. The first year we stayed in small cabins in the village and rented a sixteen-foot Boston Whaler powered by a 40-horsepower Evinrude engine that Gil had the good sense to strip clean and reassemble before we launched ourselves toward a group of rocks situated twelve miles due west in the Gulf of Papagayo. Close to the border of Nicaragua, it was the last suggestion of land before Australia. Carrying two tanks of gas, and taking advantage of the calmer conditions and shorter route, we left early each morning and cut straight across the Gulf. For a week, in front of a backdrop of jungle and white beaches, we

cast flies at roosterfish on the hunt inside the curls of breaking waves and at sailfish in the deep, blue waters of the Pacific
Ocean. When our gas ran low and the onshore winds picked
up, we made our way back to the cabins, slowly, ahead of the
afternoon storms.

The second year we chartered a sailboat owned by an expa-
triate American couple older than Methuselah and spent a
week anchored in a bay in the northern section of the Gulf. The
thirty-eight-foot schooner named *Doubloon* was as well worn as
her owners, but tethered to her stern the same Boston Whaler
we had used the year before took us to a dozen separate beaches
and to the outcropping of rocks that rose out of the dark waters
of the Gulf, even to a freshwater river that flowed down from
the mountains to the sea. In that river we waded and fly fished
knowing we were the first to ever do so.

We caught and ate corvina (sea trout) and snappers, fought
roosterfish (some on fly, most on plugs), and jumped Pacific
sailfish off the mainland, catching a few that exceeded one hundred pounds (all on flies). The sailfish were magnificent, long in
body and thick through the shoulders. Brilliant fish that came
to a fly with sail extended and bill fencing. Once hooked the
sails would jump fifteen feet out of the water, black silhouettes
against a shoreline of rocks and jungle.

To lure them within casting distance, we trolled a pair of
hookless seven-inch rubber squid a hundred feet behind the
boat on separate lines. When a bill or a shadow materialized
behind one of the squid, the second one was retrieved and the
sailfish was teased by swimming the bait in front of its bill and
pulling it away when the fish opened its mouth. Sometimes we
allowed it to partially swallow the teaser before drawing the
squid from its throat. When the bright yellow flank stripes of
the sailfish surfaced and it swam hard (Pacific sailfish can swim
up to sixty miles an hour) from one side of the boat's wake to

the other, slashing at a bait that wouldn't die, we cut the engine, jerked the squid out of its reach, and made the cast. The heavily tied white-feather fly, spun onto a 4/0 hook, was set upon directly.

We used stiff eleven-weight fiberglass rods and Fin-Nor reels, twelve-pound test and sixty-pound hard monofilament shock tippet to keep the billfish from abrading the line. Teasing billfish was as much fun as fighting them. Once the rubber squid was in their mouths, they did not want to release it, much less lose it, and that made for bill strikes and boils that split open pieces of the Pacific Ocean.

Twelve-pound monofilament made for a good fight. The fish were big and powerful, and the old photographs we took show purple-colored sailfish high in the sky, stretched horizontal, and parallel to the water. They were the sapphires in the fishing crown of Central America, fish we never killed willingly but sometimes lost to sharks.

We fished for them on the far edge of the Gulf of Papagayo, where a single hiccup in the engine would have seen us adrift across the Pacific. Such odds felt normal to us. We fished in a paradise that young anglers these days only dream of. The three of us lived an adventure no one else in the world was living. We didn't concern ourselves with consequences.

Most of my offshore fishing, though, was pursued off the east coast of Florida. There was very little boat traffic then, and no pollution. At certain times of the year, huge schools of mackerel and bluefish migrated through its inlets into the waterways. As a kid I remember filling my first outboard—a ten-foot aluminum boat powered by a 9-horsepower engine—to the gunnels with Spanish mackerel and almost sinking it in Lake Worth before I could get to the man on the dock who bought the fish for pennies a pound.

Narrow inlets such as Jupiter and Boca Raton compressed the waves, causing them to clamber up into clashes of energy that ricocheted off the rocks and cement jetties on either side of the passageways. In times of bad weather, a run through either of those inlets was unwise.

During the winter months, I fished out of Fort Pierce for sailfish that would migrate south from the Carolinas ahead of the cold fronts. Deep and wide and well marked, that particular inlet could, with the alliance of an incoming tide and a following wind, look like a monster. But it wasn't.

The return trip through the inlet after a morning of fishing required a stop outside its mouth to assess the size and rhythm of the waves, waves that had begun their long ride into the man-made pass hundreds of yards offshore. As they entered the confines of the jetties, they would rise up into monuments, smooth and set apart like a woman's breasts. While we idled there waiting for the right moment to enter, those passengers who had never made the run before, including half a dozen high-profile anglers I had fished with over the years, would beg me not to try it. As if we had a choice.

In fact, running the Fort Pierce inlet was a form of surfing and simply a matter of picking a wave, motoring up its back just short of the curl, and then riding it a hundred or so yards to calm water. Power management kept the boat from either falling off the wave ass first, or tipping forward over its lip into the trough and being chased by a mountain of water.

In 1960 Johnny Rybovich and his partner, Jimmy Dean, teamed up to make a small, inboard center-console sportfishing boat. Rybovich was a prestigious boatbuilder whose seagoing crafts were coveted by the elite bill-fishermen of the day. The lines of his yachts were known for their elegance. The boat I bought from them was an experiment, a passing fancy, an *amuse-bouche* creation for a small coterie of hard-core light-tackle anglers.

Rybovich and Dean built four prototype open fishermen boats before returning to more lucrative designs. The boats were twenty-eight feet long, V-hulled, and propelled by a single inboard screw. The propeller was driven by a metal chain, similar but heavier than a bicycle chain. When I bought the last of the prototypes, secondhand, some years after it was built, its chain had acquired the bad habit of jumping the sprockets whenever the gears were moved repeatedly back and forth—a problem when fighting a fish and a nightmare in heavy seas. Otherwise, the boat was a marvel.

The design was modern, with a center console, hatches that opened into a storage area in the bow, a bait well in the stern, lightweight removable outriggers, four rod holders, and a deck free of cleats. In short, it had everything a fly fisherman could ask for, and it anticipated by fifteen years the standard layout for modern boats of that size. Originally Dean and Rybovich included a fighting chair in the stern, but since it caught the free flow of the fly line, I removed it.

I named the boat *Streamer*, and I loved her. We fished the waters off the east coast of Florida together for three years. With hindsight, and along with a loaded 1971 Pontiac GTO and a 1978 Cessna 185 taildragger, I regret to this day having sold her.

The *Streamer* was created for blue water, and that's where I took her as often as possible. In the summer we would run offshore looking for dolphin along the long, thin braids of free-floating sargassum that at intervals collected into wide golden carpets of seaweed drifting on the surface. When a school came crashing at the baits, I would hook the first dolphin on conventional tackle, leave the fish in the water as an attractor, and chum the rest of the school into range of my fly. When the fishing was slow and the sun hot, I would jump in the ocean and snorkel under the sargassum, captivated by the world of miniature pelagic fish I would find nestled in and around the seaweed, content to live in a dense world of migrating food and cover.

In the fall the *Streamer* and I would drift offshore from the Palm Beach inlet out to a reef historically known to draw kingfish. The commercial fishermen trolled silver spoons dragged below the surface by planers. The crew handlined the kingfish back to the boat without ever slowing down. Most of the commercial catch were fish that weighed less than fifteen pounds and were referred to as "snakes."

I would blind cast into the ocean for hours, allowing my high-density line to sink the fly the length of the longest cast possible before retrieving it as fast as I could. Compared to high-speed trolling, stripping flies was slow, and I didn't get many hookups. But when I did, the king would do its best to run the backing off my reel, and I had to crank up the boat so as not to be spooled. Kingfish grow to weights exceeding sixty pounds and at that size are known as smokers. Considering all the time I spent on the water casting, I might have hooked a smoker or two, but I doubt it; if I ever did, I lost it. Fifteen pounds is the largest kingfish I remember gaffing.

Other times I nudged the boat's bow closer inshore and, switching to floating lines, fly fished schools of mackerel and bluefish—spinner sharks in the spring; snook in the summer; and tarpon, jacks, sharks, and more snook during the fall mullet run. My friends and I chummed for anything we could lure to the stern of the boat, sometimes amberjacks and jack crevalle, and once a pair of thirty-pound-plus African pompanos that we caught. The key quarry for winter fishing with the *Streamer*, though, was Atlantic sailfish, teased behind her stern and caught on fly.

The bow of the *Streamer* rose and fell to the advance of the waves spilling into the mouth of the Fort Pierce inlet. The run toward the ocean was defined by dark gray cuffs of water, wind-blown at the summit, green and washed out at the bottom. At the top of the wave, only the bow and a sky full of clouds—winter clouds traveling to the mainland—were visible. We had run north up the inland waterway from West Palm Beach. It was ten a.m. We were late.

The outgoing tide fought predominant northeast winds, causing havoc in the confines of the inlet. The wind sheared the swells at their apex and hurled spray and green water into our faces. All the boats ahead and behind us bore east and climbed waves. A cold front had urged schools of migrating Atlantic sailfish to head south from South Carolina. Sightings were high.

Roger, my fishing companion, a small agile man with black hair and a quick smile, was familiar with boats and weather. He stood to one side of the center console in a slicker, looking over the breakers into the distance for bird activity and bait. Neither of us wore hats. The bow crested the wave, dropped into the trough, and rose again. Half a dozen more breakers and we would be free of the inlet.

A few months earlier I almost hired Roger to start the first offshore fly-fishing charter company. But after witnessing the reaction of some of the fly fishermen I guided—who complained about the weather, the discomfort of an open boat, and the lack of a potty, not to mention the terror in their eyes during the inlet runs—I gave it up. The sport was still young. I was too early.

After topping the last wave, we left the rough waters of the inlet and ran east-northeast about ten miles, staying short of the blue water. Sailfish migrating south to warmer temperatures don't fight the northbound current of the Gulf Stream. I turned on the radio. Many like us were running. Some were fishing.

The winds were out of the northeast at twelve knots, the temperature fifty-five degrees and rising. It had been cold during the night. Other boats reported sightings and strikes. As we grew closer to the main flotilla of sportfishing boats, I noticed the small triangular flags hoisted on the stays of many of their outriggers. White flags proclaimed sailfish releases, red flags kills.

Minutes later I slowed down while Roger pulled tackle out from under the gunnels and bow compartment. He knew the drill. We had done this before.

I turned the boat south and trolled on a heading that took us one hundred yards west of the Gulf Stream. The same rubber squid I had used in Costa Rican waters were deployed with an additional, much larger squid strung just behind the boil of the engine to serve as an attractor. The water was green brown and dirty. A hundred yards to our left, the Gulf Stream ran indigo blue and free of the accrual of inshore waste.

After a while the wind dropped and the voices on the radio complained of a decline in the action. The waves had flattened, the sea was too calm. The squid lost their action and dragged behind the boat like mops. An hour later we reeled them in and started back to the inlet.

Roger spotted the bird activity almost immediately. Not far out in the Gulf Stream, maybe six gulls and three terns were holding in a funnel-shaped pattern above something moving inside the deep blue color of the ocean. We approached slowly from the north. When the birds did not disperse, I cut the engine and we drifted. A shimmering silver ball, eight feet across its middle, alive and terrified, was churning in the water under the birds. The ball of bait was made up of Spanish sardines. Surrounding it, five Atlantic sailfish were swimming counterclockwise in an increasingly tight pattern, effectively compacting the small, silver baitfish into a dense meal. Before I could retrieve the fly rod out from under the gunnels, and as if on cue, the sailfish ripped through the bait, beating the sardines with their bills and then swallowing the stunned ones. Once the ball lost its compactness, the sailfish—dark and dancer thin—deployed their sails and resumed herding the bait all over again.

The wind had vanished, and the bow of the boat rose and fell in the oily current of the Stream like a cradle. My cast traveled sixty feet in the air before dropping the fly—one of the white-feathered flies I had used in Costa Rica—into the terrified bait. I stripped the fly, which tempted one of the sailfish out of its pressing obsession, and a dark, conical mouth opened. The fly disappeared, and seconds later the sailfish was tail-walking across the Gulf Stream, its snaky body dancing, its bill reaching for the sky. Less than half the size of its Pacific cousin, the fish nevertheless dragged line off the reel. We followed in the *Streamer* and eventually billed the fish.

That morning Roger and I caught and released three free-swimming Atlantic sailfish on fly. The run home, through the inlet and later though the Intracoastal Waterway, was joyous. In those days the high fives and fist bumps that define today's modern sportsmen did not exist. Once inside the inlet, we shook hands.

And we grinned all the way back to the dock.

WINTER POND

From inside a jet plane at thirty thousand feet, the rivers and ponds of northern Florida occupy an otherwise bland December canvas that would have delighted the Spanish painter Joan Miró. From the airliner's porthole I observe the barren hardwoods that embrace the coils of rivers, each restrained and supported by aprons of white sand that resemble the underbellies of snakes. A uniformity of ash-gray light lends austerity to the landscape. The warm temperatures of the water fight to influence the cold air, and I see pockets of mist rising five miles below me.

Winter is a good time for putting in order the accumulation of plastic worms and plugs and jigs that have been sitting as a spring jubilee on the mantelpiece—an absurd tangle of hooks, deer hair, plastic lures, and monofilament. I have enough stuff in the pond house to fish with until I die. In the spring I'll untangle a hundred more flies, collected over time.

In northern Florida it takes a long time for winter to establish its dominion over fall. The new season approaches and retreats, returns aggressively, and changes its mind again. One day it finally settles on the clay hills and lays down a coat of frost. The fall flowers no longer cast a hue over the landscape, the blood-colored sweet gum leaves rest on the ground.

From the rocking chair I watch morning fog lift slowly from the pond through the cold air of winter in wraithlike shapes

that convey beauty, privacy, and unease. Before lunch the fog
dissipates, uncovering an icy blue winter light.

The bass live in deep water now, next to the dam, and carry
the weight of the pond on their shoulders. On warm afternoons
they rise to harass the shad. Bats emerge at sunset. At night an
occasional south wind encourages succulent memories of summer, and the peepers peep again.

On overcast days the pond takes on shades of silver, rum-
pled in spots, slick in others. The reptiles, whose hearts beat
in harmony with the falling thermometer, are land bound,
coiled inside gopher holes or under the stumps of rotting trees.
Through binoculars I await the arrival of diving ducks.

Slate-colored water shuffles under the uncertainty of a
swirling wind coerced, delayed, and redirected by the loblolly
pines that blanket the shoreline. The pond slowly turns into a
cold, gray womb at the bottom of which its inhabitants seek
warmth at depth, their appetites and metabolism measured.
Gusts of wind tarnish the reflection of passing birds. Flocks
of sparrows bank in formation as a single wing in winter. The
bald cypresses that in November painted the edges of the pond
golden are now skeletal.

Over the past twenty-five years the pond has adopted a personal
and seasonal rhythm of its own. There is heat and cold to be
expected, sometimes drought, sometimes floods. While there
is little I can do to help, I worry about the pond's well-being.
Because I understand that I am not the center of its universe, I
try to limit my concerns—more often than not ill conceived—
to the seasonal ebb and flow of events I have come to expect.

Around the middle of every January, I wait for the arrival
of the ospreys. This year, as every year, I recognize from a distance the cadenced wing beat of the male before I see its white
crown. The bird flies cautiously over the pond, its long, narrow

wings bowed downward, looking for a meal. Eventually it glides over the water to last year's nest, a wooden platform nailed to one end of a telephone pole driven into the mud.

Over the next week, awaiting his mate's arrival, the male remodels the nest by weaving in new twigs and branches and a cushion of trailing Spanish moss. Ospreys mate for life, so when his bride appears on the scene, the expectant male sky dances above his work in a proud display of expectation and pleasure. Following a renewed period of courtship, the smaller male takes over the duties of hunting and delivering food to his brooding spouse on the nest. If all goes according to nature, she will deliver between two and four eggs, which will hatch forty days later.

The bald-headed nestlings will grow quickly, and, on the rare occasions they are not screaming for food, I'll catch glimpses through my binoculars of the young birds playing hide-and-seek inside the woven assembly of their home. Once their feathers have developed, they parade on the rim of their nest, yawning open their wings for exercise and embodying the usual apprehensions of youth, hesitant to take the first jump.

Two months later the fledglings, taking advantage of the well-stocked pond, will begin to hunt fish with clumsy determination. They'll gradually sharpen their aim, until one day I'll see one of them carrying dinner in its talons. An osprey's fishing technique is not graceful, but grit and resilience are easily observed in the bird's ungainly plunge for sustenance.

During medieval times it was believed that fish were so mesmerized by the osprey that they would turn belly-up in forgone surrender. I can attest to the inaccuracy of this notion.

Early in February, in advance of the breeding season, a flock of ten or twelve Canada geese make an investigative flight over the pond and, satisfied with the size and perimeters of the habitat, a brace of birds remains. Over the next few weeks I watch the male fight off would-be trespassers, a raucous open-wing affair involving beaks and elbow blows to the chest and neck of the intruder. One year the geese took umbrage at a band of cormorants that had chosen the pond as a mess hall, saving me the trouble of calling Bill or loading the gun.

Largemouth bass in North Florida breed during the second full moon of the new year. Using their fins, the males fan a small area of mud and sand next to the shoreline and fashion a bed, a repository for the females to lay their eggs. As soon as they are ripe, the hens are escorted to the nests by the males, who in their impatience are known to bump the ventral region of their mates in order to trigger the release of up to forty thousand eggs. The males then fertilize the eggs with their milt, and when the larvae hatch (two to five days later), they guard the eggs for a few days, until hunger gets the better of their conscience. Then the buck bass joins every other fish in the pond and prey upon their progeny.

The larvae, or fry, move to vegetation that occupies the shallows and spend the next few months taking refuge from predators and feeding on zooplankton and small aquatic invertebrates. As the bass grow larger, so does the food they hunt, until they consume anything that fits in their mouths—baby shrimp, water bugs, insects, and so on. Once a largemouth bass reaches a weight of 2 pounds (in about three years), and depending on the quality of its food source, it can grow up to a pound a year. A mature male might reach 7 pounds, while a female bass raised in the South might grow to 25 pounds. Under the right scenario bass live fifteen years or so, with upward exceptions that would substantiate the 22.4-pound all-tackle record fish.

In March, six weeks after the geese take ownership of the pond, I see four small goslings paddling fretfully behind their mother as she glides over the surface, proud and elegant. The sudden appearance of new life is thrilling, almost alien. The weather is capricious. Rain stabs through sunshine. Wind hurries across the water, raising miniature whitecaps on the pond's surface. At night heavy clouds choke on thunder and lightning. The days grow incrementally longer.

Encouraged by a light breeze, the first cormorant of the season flies out of the pond's shadows, the tips of its wings carving small wedges of water on the surface. Cormorants swim with their yellow bills held at a slight angle to the sky, giving them the slightly haughty look of a bird that knows too much. They are protected birds that consume daily portions of fish equal to their body weight. I like them, but prefer snakebirds (anhingas) for their exaggerated shape and the fact that they are loners. During their migration, cormorants descend on the lakes of northern Florida in flocks ranging from a handful to hundreds. I once watched a cormorant dive up thirteen shad in four minutes. Do the math and you'll understand the problem they represent for a well-stocked pond. So when large flights of the fish birds land in the spring, I call Bill and he convinces the cormorants that there are safer places to visit.

My neighbors to the east burn their land in March. The smoky haze shields the sun and emits an ash and caramel aroma, pleasantly familiar to those of us who live in northern Florida. Prescribed burns in the South are engineered to thin the undergrowth and save the forests from runaway lightning strikes.

We set on fire about 50 percent of the farm every year, using the pond as a line of defense against proliferation. For a few weeks—or until a good spring rain soaks the land—charcoal-black debris cloaks the slopes. The pine trees stand in attendance over the dead landscape. Their smoldering stumps and the blackened shells of box turtles add to the subdued mood of the mornings, wrapped as they frequently are in fog.

If it rains I see green shoots stretch through the background of burnt earth within days. The terra-cotta tops of pine trees, singed where the fire ran hot, push out new growth. Green buds drip off the pencil-thin branches of the willow trees, and the dogwoods show white against the backdrop of spring.

Soon, as is always the case in nature, all is forgotten. The charred bodies of the victims return to earth, the foliage is restored to its former elegance, and the pine trees, compelled by turbulent spring winds, lean into their shadows.

HUNTING STONES

Looking for spearheads in a harrowed field after a rain reminds me of hunting morel mushrooms in Michigan in May. There are days when flint and morel sparkle their presence, other days when they loiter in the shadows of a tree or inside a curl of plowed earth, absent to the eye. I have hunted Indian artifacts from the Mississippian culture in the dove field above the cabin that overlooks the pond twice a year for twenty-five years. Each time Bill Poppell drops his plows into the ground and turns the dirt upside down, I wait for rain. When it has passed and the terrain is fallow, I walk up the slope to the long, narrow field and search for history.

"They glow," Bill said last spring. "Saw one this morning on the overlap of the blades, shining just as pretty as an Easter bonnet." He sighed. "I couldn't stop the tractor in time. Maybe I'll turn her over again next year."

The dove field, which is the length of two football fields, is harrowed twice a year, in January after the hunting season has ended and again in July. Bill plants winter wheat in February to provide cover and food for the quail, the turkeys, and those birds that migrate over the farm in the spring. Later he sows brown-top millet as a base crop to entice doves into the range of guns in October. For two weeks or so after Bill harrows, and before the new crop emerges, the ground is dark and orderly.

I walk in the ruts between the mounds of soft dirt and look past the dead leaves of the live oak trees that frame the dove field. Most of the shards and arrows and pieces of clay pottery I find are on the cheekbones of the field north of the pond, next to a big, petrified oak tree whose extended white arms invite blackbirds, kestrels, and occasionally a bald eagle to perch. Sun-bleached oak leaves rest on the earth. Their oblong shapes resemble arrowheads and confuse the search.

I still love to walk barefoot in freshly plowed earth. A friend said to me once, "I'm old and ugly enough not to care," but the truth is that mud between my toes feels just as good as it did sixty years ago.

In addition to the Indian artifacts I find when I search the field, I also turn up shotgun shells, pieces of black rubber tubing used to channel water to the meadows that once housed white Charolais cattle, the remains of barbed-wire fencing, and modern bits of PVC tubing.

The native pottery I find is always broken, but it is recognizable because of the cloverleaf-shaped designs that surround the mouth of the vessels. The spear points are tapered, dimpled, and sharp. Shaved vertically to accept the shaft of a spear, the points often display an oblong depression in the flint, which I stroke as its maker did five hundred years ago.

What we as a country did to the Native Americans, owners of the land we coveted and now take for granted, is a continuing abomination, in my mind, as bad as or even worse than our treatment of the pre–Civil War slaves imported from Africa. Beliefs and delusions are incestuous brothers. Our behavior toward both races reflects the genuine inhumanity of the human species.

Tony Arnold, my paleontologist friend at Florida State University, had the artifacts I found in my dove field carbon-dated. The spear points had been fashioned a few decades before Hernando de Soto started his homicidal march up the Florida peninsula in 1539. A year later he commanded his army across the Ochlockonee River north of Tallahassee and marched toward Bainbridge, Georgia, very close to or perhaps even over the farmland I live on. The pottery pieces were created much earlier, around AD 800, which implies that Indians from the Kolomoki chiefdom fashioned them. The Kolomoki were eventually replaced by the Mississippian Indians, who in the time of de Soto were known as the Apalachee, Chacatos, Amacanos (on the coast), and Apalachicola tribes. In the eighteenth century these tribes were united under the name of Creek.

The gap of time between the pottery and the "points" that I find next to the petrified live oak suggests that generations of Native Americans lived on the same hill. There they chipped flint and shaped bowls for eight hundred years while keeping an eye on the swamp that now invites my gaze as a twenty-seven-acre pond.

"Don't brag about it," Tony warns me, "unless you want people running all over your land claiming state rights."

Accepting Tony's compression of geological time (4.5 billion years into a single year), the first living things (the common ancestors) appeared in the sea in May. Land plants and animals appeared in late November. Dinosaurs dominated by mid-December but disappeared on the 26th, about the time the Rocky Mountains were uplifted. Manlike creatures emerged during the evening of December 31st, and the continental ice sheets began to recede from the Great Lakes about one minute and fifteen seconds before midnight. Rome ruled the Western world for five seconds from 11:59:45 to 11:59:50. Slavery and the purging of American natives to reservations came and went one second before today's Lilliputian year of years.

A sobering thought when one realizes that all the harm we have inflicted on the planet since then has been carried out in the blink of a geological second.

The buffer of consciousness we hide behind, a trait allegedly unique to Homo sapiens, is porous, its permeability increasing with the growing numbers of our kind and the percentage who live in poverty. Since the planet is equipped to handle at best two billion people, and we are racing toward eight, it is not surprising that nature is sending out reminders to us in the form of floods and droughts, famine, and disease-driven depopulation. It is impossible to keep adding numbers to a shrinking environment without consequences. The biology won't allow it.

Each outing into the plowed field reminds me of the excitement of a first date. I never know how it will end. Most times I spot only shards and slivers of flint, sometimes a point vexingly split in half or broken at the tip. But sometimes, perhaps twenty times in all these years, I bend over and pull from the ground something breathtakingly beautiful, something as long as a child's thumb, usually striated brown with yellow veins, occasionally white, perfectly preserved since the days it was created half a millennium ago. Three times I have plucked out of the earth a seamless spear point, lovely and pink and precious as a pearl. The conch pearls of Gadsden County.

Because there is something magical about arrows and spearheads fashioned centuries before our European ancestors even dreamed about this continent, tools for hunting, carved by a culture we eventually decimated, for a long time were called "elfstones."

TOM AND THE FAT BOYS

Over coffee one day during the spring of 1969, Woody Sexton introduced me to his friend Tom McGuane. At the time, Tom; his wife, Becky; and their young son, Thomas, were living on Summerland Key. Tom, a tall, handsome writer in his late twenties, was fishing the Loggerhead Basin from inside a sixteen-foot Roberts skiff powered by a 33-horsepower Evinrude engine and steered by a tiller. That winter he wrote his first piece for *Sports Illustrated* magazine, "The Longest Silence," maybe the best article ever written on flats fishing. After reading it that summer, I sent him a congratulatory letter, and he responded by inviting me to join him that fall on his ranch outside Livingston, Montana, for a week of trout fishing. To this day I am surprised I made the trip, because it has never been my habit to accept sporting invitations from people I don't know well. In this case, and in retrospect, I am grateful that I did. We have been close friends ever since.

When the enthusiasm of an angler toward the wonder, order, and harmony of nature takes precedence over the numbers and size of his catch—and if it is the angler's choice to mostly fish alone—the experience develops into a form of meditation. It was clear from the *Sports Illustrated* article that Tom was in full harmony with the natural environment he was describing. Whether it was sunlight transitioning over sandbars, the diligence of shorebirds feeding, or the collective terror of a school of mullet, he treated the everyday instances of nature with the same thoughtfulness with which he treated the pursuit of fish. To this day Tom hunts and fishes for the pleasure of being in the field or on the water, usually by himself with his thoughts and observations, which he considers evenly with the action at hand. In the case of "The Longest Silence," it was catching a permit on fly.

On my first trip west, Tom took me to all the lovely, tranquil spring creeks around Livingston and Bozeman, and of course to the Yellowstone River, where late one afternoon he caught two five-pound brown trout on consecutive casts. It was a big deal. They were "wall" fish and treated as such.

Coming from salt water, I considered a five-pound any-thing to be a fish you ate or released. As well, I didn't think the trout put up much of a fight, but there was no denying that they were beautiful—slick and brown, blood spotted, and yellow bellied. The sight of them clean out of the water took me back to my childhood in France and to the confluence of two rivers.

None of my close friends dwell in cities, and we all have deep relationships with animals, mostly with our dogs, some of us with birds, others with fish, Tom with his horses. Mostly artists of one sort or another, they all reside in rural communities or by the sea, where nature persists as the prevailing influence on their lives. Their art, whether it be poetry or prose, painting or music, is a reflection of the environment they choose to live in.

Tom invited me back to Montana numerous times, and the more often I traveled west, the more I took pleasure in the subtlety and minutiae of dry-fly fishing. Wading in rivers as opposed to the ocean introduced me to the unfamiliar weight of moving water and imposed an understanding of depth and flow and the behavior of a dry fly sailing across a quick, shal-low run or floating high on a film of calm water. I watched flies

knife through the shade of undercut banks, perch on the surface of deep pools, and teeter past the wake of submerged logs—
pictures that are foreign to a saltwater angler.

By the middle of September, the summer sun had baked the
valleys, leaving them uniform and golden. Weeds framed bright
green fields of second-growth alfalfa that early and late in the
day filled up with mule deer and antelope. Gray partridge chased
grasshoppers across the rocky slopes of hills under the watch of
eagles and hawks going about their business below the inexorable spread of winter at higher elevations. In October the brown
trout swam up from the Yellowstone River to lay their eggs in the
clean water of Armstrong Spring Creek.

The creeks outside of Livingston and Bozeman offered classic dry-fly fishing, and in the mid-1970s they presented mini-
mal intrusion from other anglers. Hatches rose from their beds
to the surface every fall afternoon, and without any concern or
awareness of time passing, I would fully concentrate on the task
of laying a tiny dry fly quietly on the water in the right place.
Hours came and went like songs, with the conclusion of the
melody signaled by a chill that slowed and then terminated the
emergence of duns.

Intricate, vermiculated patterns, pale blue halos, red belly
fins, and bright orange stomachs drew me to the brook trout living in the spring-fed ponds above Paradise Valley. The brookies
were members of the char family and had been introduced to
Montana in the late nineteenth century. I loved to eat them,
cleaned and shaken in a paper bag with brown sugar and dry
Coleman's mustard, a recipe from Al McClane's classic cook-
book. Fried in butter, the pink-fleshed fish left a mess at the
bottom of the pan and delight on the tongue.

Inside an inner tube on Silver Creek, in Idaho, I floated
through high-desert pastures, past black cattle, and under blue
skies while watching flies that I had cast forty feet ahead of me.

Huge imitations of grasshoppers and beetles, the flies drifted at exactly the same speed at which I floated for hundreds of yards until a fish struck or monotony drove me to make a new cast. I can still see the flank of a large brown trout roll out of the water at my fly drifting next to the bank. Ten, twelve pounds? I don't know, I never felt her.

My original bias toward the size and endurance of trout came from learning to fly fish in the sea, where the horizons are limitless and the struggle to stay alive, herculean. Rivers are more contained, more feminine; the enemies of the fish that patrol them less abundant. By the end of my first trip to Montana, I dismissed the diminutive size and lesser fighting abilities of trout and simply took pleasure in the fact that I was fishing in hauntingly unfamiliar settings for fish that were particular and often difficult to entice.

Tom moved his family to Key West in 1970, and his friends followed. One of my most vivid memories of that year is of poling Tom and his college friend, poet Jim Harrison, across the flats between Mule and Archer Keys one morning in May during my first spring in Key West. I don't think either hooked a fish that day, but I remember listening to these two men talk about novels with an ease and facility that I hope I might have deployed discussing the merits of a best London shotgun. I was not familiar with a single title they discussed, even though growing up I had always been a reader (there was no television in France in the 1950s). The fact of the matter is, I had not graduated beyond the books of Ian Fleming, Rex Stout, and John D. McDonald.

It has occurred to me since then that my literary education did not begin during the eleven years I spent in boarding schools, or during my single year of college, but on that day inside a Fiber Craft skiff poling for fish on the flats west of Key West.

Consequently, just as it took me a decade to learn how to cast a fly correctly, it took me at least that long to understand and value Tom's novels and Jim's poetry.

Tom and I fished together in the spring of 1971, mostly for permit, when permit were considered almost impossible to catch on fly and long before anyone had dreamed up the modern crab-fly imitations. That year and the next, Tom and I cast at hundreds of permit—small and large committees of fish working the brittle edge of the sandbars, big pairs and singles on the face of the flats, and hundreds of black sickle-shaped tails protruding out of the water, belonging to permit feeding. We ran and fished over miles of shallow water between the Bay Keys and the Marquesas. Although we were both good casters and poled a skiff as well as any guide, the permit—which look like a cross between a pompano and a jack crevalle—did not react to our flies. The fish were as discriminating as trout during a hatch, and we never presented the right match. We fished off and on together for two years, and while we jumped tarpon and caught muttonfish, we never hooked a permit.

I remember fishing the tides between Man and Woman Keys and the rack of clouds that took shape every afternoon from the distant Marquesas to Key West. Small blacktip sharks hastened from one point of interest to the other with a purpose best suited to their single-minded brains. Lemon sharks transferred a more relaxed rhythm to the surface. Menhaden shivered across acres of deeper water, and with the advent of sunset, the clouds melted and the sun fell slowly over a river of water unfolding out of the channels onto the sand. The wind would die, and with the emergence of the moon, the flats would swell into lakes of cathedral proportion.

The next morning, ineffable and complex in their symbiosis with the rising tide, the flats would come to life. When the stingrays began to forage, Tom and I would pole toward them, searching for cormorants hunting in the cloudy disturbances created by the rhythm of the rays' wings. Shafts of light deepened and saturated the color of the sand. Popping their heads out of the water every few seconds with a crab or a shrimp held in their bills, the birds were often joined by mutton snappers, ten- to twenty-pound fish whose red tails waved out of the water when feeding. The snappers took flies well, and once on the line they blistered the flat, raising rooster tails in their wake. Troubled by the activity and also seeking safety in deeper water, the stingrays hurried after them, leaving great wing prints on the surface.

For well over a decade, Tom, Jim Harrison, the painter Russell Chatham, and I spent our springs fishing together below Key West. We poled over the alarmingly white flats between Man and Woman Keys, watching for the reflection of tarpon to rise from the sand. We cast bonefish flies at tarpon in water so shallow their fins left furrows on the surface. We fished quietly over laid-up fish in the Peal Basin, where the tarpon assumed a slight discoloration before melting into the sea grass. We jumped tarpon in the northwest channels and off the naval base. We fished named flats such as Loggerhead, south of Big Pine Key, and the Eccentrics, west of Big Torch Key. We fought fish in Mooney Harbor inside of the Marquesas, we hooked tarpon under the Seven Mile Bridge during the palolo worm hatch, and we made long casts under the night lights of the Pier House Hotel in Key West.

Tarpon, tarpon, there were tarpon everywhere.

We arrived in Key West at the end of an era and left at the beginning of a new one. In the early seventies the town was sultry, magnificent, and suspicious, the home of cockfights and unclaimed ladies, tough bars on Duval Street, knifings on the shrimp docks, and hippies partying in Mallory Square.

Hibiscus and bougainvilleas fell from the balconies, and the poinciana trees glowed like setting suns. In the morning we rose to shots of black Cuban coffee.

The landscape of the flats that shifted under the inspiration of currents and daily tides was pristine and, compared to today, empty. The Internet, Jet Skis, and cell phones had not been invented. Cruise ships docked at other ports. Advertisers had not discovered the angling apparel that ignited the fly-fishing craze or the infrastructure of hotels and guides that would be needed to accommodate those newfound anglers. The flats were still-life compositions of light. Visited only by a small commercial fishing industry, a handful of illegal netters, Cuban crabbers, and a few pot runners, the sea was healthy and the marine life plentiful. Those of us who fished the Keys in those days were the most fortunate of all anglers. The competition was nonexistent, the fish unmolested, and the flats silent and immaculate.

When we left the Keys in the early 1980s, the downtown shops on Duval Street were filling up with bric-a-brac and T-shirts, and the first of a thousand more to come Carnival boats had disgorged its load of pink-fleshed tourists with cheap memories on their minds.

Jim Harrison, Russell Chatham, and I were known in Key West as the "fat boys," as in, "The fat boys are back in town," loosely translated as, "The party is on." Not that Key West needed our encouragement to throw a party. We merely added our weight, enthusiasm, and appetites to the mix. We weren't particularly fat (certainly not by today's standards), but we all cooked and ate well.

By 1974 Tom was often absent during the prime tarpon months, busy writing screenplays and making movies for Hollywood. Through the 1970s and into the early 1980s, Jim, Russell, and I rented a house in Key West for six weeks every year and parked my skiff at Garrison Bight. Every morning, no matter how distraught we felt from the previous night's indulgences, if the weather was tolerable we fished, or attempted to fish. One morning—I have been told this but don't remember—my Maverick skiff was spotted, by a couple of guides and their anglers, drifting inside Mule and Archer Keys with the fat boys—the poet, the painter, and me—sound asleep on its floor. The biblical hangover, forgivable, since it was a weekend.

Then there was the day we drank rum and Cokes on the flats beginning at ten in the morning. It so happened that one of us jumped a fish while another was indulging in a Cuba Libre to chase the fog that had settled after a long, sleepless night. A second fish was spotted at the next mixing of drinks, and from then on our luck increased each time one of us poured rum. Tarpon swam in range of the boat at every swallow, and since it was more fun than studying the tide charts, we persisted. It was magic, and by the time the bottle was empty, we were seeing fish everywhere. The memory of our run back to Key West across Northwest Channel that afternoon surfaces with surprising clarity forty years after the fact. Mercifully it recedes just as quickly back where it belongs.

Neither Jim nor Russell was handy with the pole, so I did the pushing. Poling gave me a perspective into the world of guides and their clients—the choices that lead anglers to tarpon, and the importance of pointing out the fish and setting up the skiff for them to make the cast. The water temperature, the tide, the water level, the contour of the flats, and the adjoining channels all tell a piece of the story of shallow-water tarpon, and I soon found the hunting of these big fish and the excitement they provoked in the boat to be as entertaining as the fishing.

For years, after a day on the water, I would tie knots and flies. Over time it would be hundreds of nail knots, clinch knots, blood knots, Albright knots, and the Bimini twists for which I used my big toes to open the loops of monofilament and set the knots spinning. My two friends, the artists, pretended not to understand how to tie tarpon leaders. As insurance against having to learn how it was done, they declared that they simply "couldn't take criticism." It was the perfect foil against any and all inconveniences.

I no longer wanted to waste bar time wrapping monofilament, so I made our shock tippets using two-weight leader wire twisted at both ends through a small swivel and the eye of the fly. The leaders took twenty seconds to make. Since we were interested in jumping fish, not in records, the leader wire version worked fine. In fact it probably worked better than monofilament, given that the wire dragged the fly down to the fish faster.

Back in the days when the calendar and geography worked, Tom, Jim, Russell, and I met and took advantage of the fact that we loved books and art and dogs and birds and fish and food and good-looking women. We fished and hunted and drank and cooked from one end of the country to the other for a quarter of a century, with Key West as a beacon of our sporting year. Now when we see each other, we remember what nonsense we used to get into and how even though we thought we did, we never got away with any of it.

What I remember best about those decades was the laughter.

Every time our group was together, we laughed and laughed, often to the point of hurting. Head-splitting, belly-heaving silliness at all times of day and night, in the boat, at the bar, during dinners, in Key West, in Montana, in the Upper Peninsula of Michigan, in France. Everywhere and anywhere, we laughed and laughed and laughed, and I miss it.

The other day I was at the open market in Tallahassee and one of the vendors I know answered my query about a plant he was selling.

"It is the calyx of the hibiscus flower. You make herbal tea with it." Then he looked at me and added, "You obviously weren't a hippy, back in the day, were you?"

Before I could shut my stupid mouth, I replied, "No, but I sure woke up next to a bunch of them!"

He looked surprised and then smiled, remembering.

One afternoon Jim, Russell, and I were fishing the flats north of Boca Grande. We had jumped four tarpon that day, three between Mule and Archer Keys, and one off the Seven Sisters, with more rolling toward us. We were staked out on a point of sand overlooking the channel that separated Boca Grande from the Marquesas.

A hundred yards to the north, inside the dark, narrow channel that split the flat in two, a school of tarpon was daisy chaining: a merry-go-round of hundred-pound fish following each other in a mock breeding ritual. Big, confident, ocean-going fish.

Under normal circumstances we would have been walking into the Chart Room, a nondescript bar we frequented every evening, but on this day the weather was beautiful and we knew the tides were right for fish to swim past Platform Point. Soon the bronze head of a tarpon rose out of the slick calm water and sighed. Like no other fish, when the oxygen content of the water is low, tarpon rise to the surface and breathe into modified swim bladders, producing a ghostly sound that flats fishermen hear in their sleep.

A school of five tarpon, backlit in the waning light, appeared for an instant offshore from us and then changed directions and, for reasons of their own, split into fingers of unease. Russell stood on the bow, a big, gentle, one-eyed man wearing a mustache and a great European nose that preferred one side of his face to the other. He was, like Woody, a product of the steelhead rivers of California. His high casting motion was not as well suited to the windy sweep of the flats as it was at heaving lead core fly lines into the bodies of rivers. But because he had spent decades with a rod in his hand, Russ instinctively knew where to put the fly and how to swim it.

Jim stood behind him, smoking a cigarette and volunteering advice. Also sporting one good eye and a mustache, Jim carried a rock-hard soccer ball stomach and a sharp sense of humor.

"What did that woman mean last night when she told us that all you wanted her to do was to yank on your gherkin?"

"Jeez, Jim," Russ answered without looking back. "I'm trying to concentrate here."

When the tarpon regrouped, they resumed their travels toward the skiff. Russell raised a high, open loop of fly line and, since there was no wind to interfere with the cast, the line unfolded and settled his fly quietly on the water in front of the approaching swell shaped by the school. Russell moved the fly once, a short pull in front of the lead tarpon. The fish raised its head out of the water and heaved forward. The fly vanished. An instant later the tarpon climbed out of the water, contorted and unbridled, exulting in its reach for freedom. The low light illuminated the platinum-colored flank of the hundred-pound fish and momentarily stamped its reflection on the surface of the water.

The tarpon ran from the skiff across the pale grass toward the channel where the school had been daisy chaining earlier. Turning south toward the Marquesas, the tarpon followed the canal out to the broad flat that spilled from its mouth, and once in the Boca Grande Channel it jumped again: a miniature pendant against a setting sky. Russell tightened the drag and broke the fish off.

It is the tarpon's movements in and out of water that interest me. If the tide is right and the tarpon are running, I don't see the point in fighting them when I could be casting at fresh fish. Seeing the tarpon underwater, judging where to cast the fly, managing the strike, and witnessing the first couple of jumps is where fly fishing for tarpon begins and ends for me. A fight is a fight, and when I was younger I fought dozens of tarpon. But now I leave the manhandling to others. For me the finesse of the sport ends a hundred yards from the boat.

In my day tarpon were killed for pleasure by men who loved competition. Tarpon tournaments fed egos. Later, once the awards ceremonies and the revelry ended, the tarpon lost their status as icons and were hauled to land dumps or dragged offshore as fodder for the sharks. Some anglers revel in the techniques of combat, just as others take pleasure in lifting weights. I like speed and focus, beauty and motion, and I believe that respect is owed to every heartbeat on the planet.

It took one hundred years of killing tarpon for no reason before things began to change. In this country those kill-tournament days are over; the law forbids it. Once again tarpon are icons, but of a different sort, and their mysterious migrations are being studied by instruments as magical as those that revealed to my computer the resemblance of my pond to a bird. Part of me wants to know where the tarpon I see each spring go for the rest of the year, the route of their migration and where they breed, but just as I would want the past history of a lover to remain a mystery, a larger part of me wants this fish that I love to retain the secrecy of its existence and simply show up once a year in places he and his ancestors have called on for millennium. In this age of revelations, mystery is a valuable commodity. Progress often takes away what it took a long time to create.

SALMON

Better suited to my temperament than trout are the anadromous fish that return from salt water to the rivers in which they were born, agile and strong from surviving the sea. Although my experience with steelhead and Atlantic salmon was and is limited, for a few summers I traveled to the Orofino River in Idaho, the Restigouche River in New Brunswick, the Deveron in Scotland, and once to the Laxa in Kjos in Iceland. I cast a thousand flies on each of those rivers and caught fish, beautiful silver fish embossed with sea lice, fish that fought well and ate better. But I did not spend the decade necessary to understand the nuances of running water, the salmon lies, the pools, the use of Spey rods. Had I spent the time to learn the fisheries, I would probably have focused my efforts on the steelhead that glorify the wild rivers of the West rather than on the Atlantic salmon, but that's conjecture. They are both wonderful and enigmatic fish.

On my first trip to Canada, I cast for three days in the waters belonging to the Restigouche Salmon Club—pretty flies, single-hook flies, gaudy flies, double-hook flies, tube flies, Jock Scotts, Stoat's tails, Munro Killers, all sorts of flies—without a bite. I fished out of a comfortable twenty-seven-foot cedar canoe manned by a competent oarsman and a gillie. It was the gillie who decided when to drop the anchor, what fly to use, and when and where to cast it. The river was bordered by a dense evergreen forest, and the angling was terribly civilized. The summer weather was beautiful. I felt superfluous.

By the end of the third day, I had made twenty-five hundred casts, an uncommonly high number for a saltwater fly fisherman. As I had done every evening, after finishing on the river I returned to the lodge, Indian House, to admire my companions' catches and listen to their stories. They all commiserated with the young angler who could cast the entire fly line, nodded their sympathy, and turned their attention back to the dice dancing across the backgammon board.

The fourth and final morning found me fishing the "Million Dollar Pool," so named because at the turn of the twentieth century a rich American had offered the club one million dollars for the water. The club turned him down. I had been using my eight-weight fiberglass bonefish rod on the Restigouche, a Pflueger Medalist reel, and a ten-pound test tippet, none of which made my French Canadian gillie, Francois, happy.

In those days I could cast with both hands and therefore fished from either side of the boat without endangering the guides' hats. My team liked that, but they had never heard of the quarry I fished for on the flats, so when I proudly told them that earlier in the year I had caught a thirty-pound permit on the same rod I was using with them on the Restigouche, they inquired what the permit was for and why one was needed to fish in the sea.

First thing in the Million Dollar Pool, before the morning fog had lifted from the river, I hooked and caught a grilse (a young Atlantic salmon on its first return from the sea). Then, three casts into the following drift I hooked another, much larger salmon. Following orders barked out by Francois, the oarsman landed the boat on the rocky shoreline and the gillie helped me out of the bow to fight the big fish that had just slingshot across the pool. He wasted no time in telling me what I was doing wrong, namely, everything.

"You battle with this fish like you did the petit. You lose it!

"Why you only have ten pound tippet? It is idiot!" he added.

"You reel it is broken. Listen!" Exasperation clouded his brow.

Francois was accustomed to the murmur of Bogden reels, not the ratchet sound of a Medalist. The harangues I had been first introduced to in the Keys in 1967 had found me a decade later in Canada.

The hook-jawed male salmon hurtled through the mist and, in its determination not to be caught, spiraled over the deep, dark water of the pool. I was lucky he did not make a run for the fast water that flowed below the Million Dollar Pool to the sea. Instead the male salmon stood his ground and wrestled with me from within its confines. The reflection of the sun pushed itself across the river and burned off the last tendrils of fog. When I was finally able to coerce the big fish into shallow water, the gillie was waiting with his net and the pool was brightly sunlit.

The mature salmon, which had returned to the Restigouche River after three winters in the Atlantic Ocean, lay stretched out and subdued on top of the smooth rocks that lined the bank of the river. The fish was deep bodied and long and just as bright as the silver platter my grilse would be served on for supper at the lodge that farewell night. The gillie and the boatman were very excited. At the time I did not know what a thirty-pound salmon represented to the angling community, but I was aware that the fish I had landed was big. In those days the concept of catch-and-release was scoffed at by salmon anglers and gillies alike. They were adamant that once hooked and fought, salmon died. Francois applied his "priest" to my fish's head, and we shook hands.

And, since the Restigouche River had a two-fish daily limit, I was back at Indian House by nine o'clock in the morning, with little else to do for the rest of the day.

To me, the beauty of rivers is less subtle, less monochromatic than the beauty of the saltwater flats. My time fishing on rivers often equaled in quality my best days of fishing for permit and bonefish, and even for mutton snapper (back in the days when they hunted shrimp and crab in shallow water). But with one exception I do not remember a single day on any river in any country that equaled the thrill I felt when guided by tide and topography, I would pole up to a school of rolling tarpon, their size and intent reshaping the surface of the flat.

The exception occurred one summer morning fishing Iceland's Laxa in Kjos with Russell Chatham.

Russell and I arrived in Reykjavik early one morning with six other anglers. After the perfunctory nap taken between sheets that—just like the hotel toilet paper—did not profess to be soft, we sat down to a late lunch at which the maître d' tried unsuccessfully to interest us in tasting some of Iceland's national dishes, such as blood pudding, sheep's head, and fermented shark. Later that afternoon, as a group, we took a walk to the tackle store where for a fortune one could buy a variety of flies— named and familiar to every salmon fisherman in the world—at twice the cost one could buy them in the United States. A pretty salesgirl picked out wool-and-angora Icelandic hats for everyone. She was lovely, with short wheat-blond hair, blue eyes, and a face innocently receptive to Nordic fairy tales. When she raised her arms to place the hats on our heads, she might as well have been selling us stovepipes. We were all in love.

Russell asked where we should go to have a drink later that evening. She wrinkled her little nose and gave us the name of a couple of bars. Then she frowned and said, "But, it is Wednesday!" She explained that no one in Iceland goes to bars on weekdays, that the clubs she mentioned would be empty until Friday.

That night our table of anglers shared delicious platters of cold-water langoustines and, as a main course, sea grass–fed lamb. None of us had fished Laxa in Kjos before, and after a few beverages the members of the team began arguing the merits of various salmon flies. Russ and I chose not to partake in the discussion—Russell because he knew a great deal about flies, having cast them at thousands of anadromous fish on the west coast of the United States, and me because I knew nothing about them.

Russell had warned me not to pay attention to the lengthy debates on salmon flies, which he accurately predicted would occur every day. "These guys discuss rods and reels and flies like we might break down a complicated cooking recipe, except the conversations are much longer. Some of them talk more than they fish. Don't worry. I'm bringing a vice grip and I'll tie the flies we need." He had suggested I bring a seven-weight rod and a floating line, and that he would make up some twenty-five-foot-long high-density shooting heads in case we had to fish deep.

I knew not to concern myself with the details. Russell's passion was steelhead fishing, and he had spent as much time chest high in cold river water as I had spent knee-deep wading the warm flats of the Bahamas. Both salmon and steelhead belong to the same salmonidae family, and once they are back in the rivers of their childhood, they behave similarly. Neither Russell nor I cared what equipment we used to cast our flies as long as the fish ate them. Gear was not a subject we indulged in. Making the choice between reading *Field & Stream* or *Penthouse* was obvious to both of us.

The Laxa in Kjos is a medium-size river at the bottom of a glacially forged valley located one hour northeast of Reykjavik.

The first fishing beat starts a few hundred yards from the sea and receives the freshest salmon of the year. The remaining fishing water tumbles over waterfalls and high escarpments, through a broad valley with wide bends and meadow pools, and finishes upriver in a narrow canyon where the flow is quick and the water deep.

The lodge, a nondescript building that served as a lounge and dining area, was flanked by four smaller houses, each containing two bedrooms, all of which were clean and stark without a single picture on the walls. Bad news greeted us shortly after our arrival. "The salmon haven't moved into the river yet," Yorgie, the manager, told us at the check-in desk. Fighting his Nordic predilection for gloom he added: "But we expect them here any day."

We sat down for lunch sobered by the report and were served the first of many "white" meals: white soup, white bread, white fish (poached), creamed potatoes, white asparagus, and steamed cauliflower, which engendered a bleached interpretation of the maxim: "Eating, a prelude for shitting." A tall, thin girl with short, curly black hair and disturbingly pallid skin served the meal. She paid special attention to Russell who, as he always does around women, smiled at her. Her name was Oliwia but we called her Olive Oil, after Popeye's girlfriend.

Later that afternoon, after readying our gear and trying to be positive about the upcoming week, we pulled on our waders, laced up our boots, and marched single file from the lodge to four waiting cars belonging to our guides. The guide assigned to us was Donald, a young, handsome Icelander. He worked winters in a sporting-goods store in the nearby town of Kerlingar-skard and spent his summers guiding. Donald confirmed what the manager had said about the absence of salmon in the river.

He took us to a series of pools and bends in a beautiful valley under a sky that every hour adjusted its structure and

hue effortlessly, from an open-eyed marvel of silver on water
through all the shades of pinks and oranges to the blood colors
cast when the sun begrudgingly fell below an uprising of dark
clouds. Since it was summer in Iceland, darkness arrived and
retreated between one and four in the morning. The rest of the
time, ambient light permeated the landscape. Before returning
to the lodge, Russell and Donald and I walked along the banks
of our meadow beat and only counted four fish spread over half
a kilometer of clear water.

One of the guests in our party had lost a grilse and caught
a salmon. Russell had felt a bump at the end of his line, but that
was all the action our group could boast of.

We had a visit before dinner from Bergulfur, the owner of
the local fish-processing plant. He was taking orders for the
quantity of salmon we wanted to take home: smoked or made
into Icelandic gravlax (raw salmon cold cured in salt, sugar, and
dill). Four fresh salmon would yield one processed fish. If the
angler did not reach the quota, there would be an extra charge.
If the angler turned in more salmon than the number of fish he
ordered to take back to the States, he would receive a check for
the difference. Bergulfur's timing, perfected after years of deal-
ing with wealthy, middle-aged American anglers, was faultless.
He took the fish orders forty-five minutes into the first night's
cocktail hour.

The guest who had caught the only fish of the day got hor-
ribly drunk that night, and he and his roommate missed most
of the next morning's fishing. After dinner Russell assembled
his vice grip and tied his version of a Comet fly—orange-
and-silver feathers tied sparsely onto a number-four Mustad
hook—a fly he had used on steelhead for years. We would be
fishing the first beat of the river in the morning, and since the
water was fairly high and the fish scarce, Russ insisted we each
use one of his high-density shooting heads. The heads were

twenty-five feet long and nail knotted to sixty feet of thirty-pound monofilament.

The next morning Donald drove us to our assigned area: the first beat, a hundred yards inland on the Laxa in Kjos. When we reached the bridge that crossed the river, we could see the bay, ruffled by an onshore breeze. Farther out, the deep blue canvas of the Atlantic Ocean was filled with breaking waves. On the inland side of the bridge, the Laxa in Kjos, wide and shallow at that point, concluded its run to the sea by skirting a minefield of round boulders protruding out of the water like otter heads. Farther up the valley, beginning with beat number two, deep swirling pools were gathered below a series of waterfalls the so-far-nonexistent salmon would have to scale. A shallow band of mist drifted over the river. Solitary clouds rode the wind and painted quick intrusions of shade on the surface of the water.

The first salmon jumped while we were walking down to the river from the knoll next to which Donald had parked the car.

"Did you see that?" Russ said. He quickened his gait. A second fish jumped farther upstream, followed by another, and another.

"Holy shit!" Russell was running.

And then we were in the river casting and laughing and pointing at bright silver fish leaping into the current of the river they had returned to. Hundreds of salmon were streaming out of the ocean into the Laxa in Kjos, and we were where every fly fisherman dreams to be, ahead of them.

The shooting head Russell had made up was perfectly matched to my seven-weight rod, making the casts effortless. Soon I was vicariously following the ride of my fly in the current, underwater, past a smooth round stone sixty feet into the river. When the fly reached the quiet water behind the boulder,

I felt a familiar weight and raised the rod. The feel of the first fish of the day is always a revelation.

Soon Donald was running back and forth between Russell and me, scooping up salmon and whacking them on the head. The salmon, clean and bright, fought with the tenacity of their saltwater peers. They weighed between eight and twelve pounds, and each time either Russell or I hooked up, they ran us downstream. With fish running upriver everywhere, hooking them was a matter of casting into the deeper current flowing on the edge of the gravel flats, or in front of a boulder.

The salmon were moving upriver on a conveyor belt, and after a while I knew to stand on a shallow rock I had marked in the river and cast sixty feet toward the far bank, allowing the monofilament to run through my fingers until I felt the piece of yarn I had tied thirty-five feet back from the fly line. When I pinched the knot to end the cast, the fly landed in what Russ referred to as "the meat bucket," and with it came the strike.

Donald laid each salmon on the grass behind us in rows of five. Halfway into our allotted twenty fish, I looked up to see Olive Oil standing next to a busload of Japanese tourists who had disembarked on the bridge and were pointing at us and jabbering. I saw her wave at Russell. He waved back.

God only knows how many salmon we could have landed that morning had we been allowed to keep them all, but once we realized we would fill our quota, we played games with the fish by not setting the hook, by casting at the shape of fish moving upriver in very shallow water, by stripping the fly as one would while barracuda fishing in the Keys, and eventually by watching each other fish. Russ caught more salmon than I that morning, a pattern that continued on every beat of the Laxa in Kjos during our five days of fishing in Iceland. At times I would climb up to the top of the waterfalls and, from high above, watch my friend wade into the current until he was inches from filling his waders. He would raise his right arm and make long, beautiful casts into pools and eddies that no one else could reach from his side of the river. Russell's fly rod bowed through space as deliberately as if he were stroking the string of a cello, and he delivered a symphony of casts as extreme as the projections of a Stradivarius. Russell Chatham was the Yo-Yo Ma of the Laxa in Kjos.

At lunch that day, after we had accepted our genuine and begrudged praises, we recognized that our next beat was number eight, situated at the farthest end of the river. Our fishing hours were divided into two six-hour shifts, with the evening's activities beginning late in the afternoon and ending between ten and eleven.

"There is no way those fish will make it up that far by this afternoon," Russell said.

"What do we do?"

"It's Friday. Let's go to Reykjavik!"

That evening we ate more langoustines and lamb at the hotel and then took a nap. We planned to be at the club the pretty salesgirl favored around midnight, an approach to nightclubbing we had perfected over the years in other big cities. The idea is to walk in clean and sober—beacons of light—in an otherwise drunken melee.

Little did we know! The nightclub was big and square and loud with posters on the walls and flashing lights synchronized to loud disco music that repeated itself every forty-five minutes beginning with the song "Staying Alive" and ending with "*Voulez-vous coucher avec moi*"! And, in the middle of the huge dance floor, two hundred of the most beautiful girls I had ever laid eyes on were laughing and dancing, mostly with each other.

"Sweet Jesus," Russell exclaimed, holding on to my shoulder.

It was a wave of blond hair, blue eyes, and milk-white skin, open faces, singing voices, and short dresses, girls with their hands in the air keeping in tempo with the music. As we made our way to the closest bar, some of the girls waved at us, as if we were old friends; others giggled and whispered things to each other we couldn't hear. We walked through a three-deep huddle of men next to the bar and ordered drinks.

The men were young and fit and very drunk. In fact they were so drunk that some of them were already sitting on the floor. We were witnessing Nordic binge drinking, except instead of beer, the lads were downing shots of aquavit and vodka as fast as the bartender could fill their glasses. It was frightening, and according to the girls who wandered over to talk to us, it was normal and to be expected. On weekends the men got commode-hugging drunk, but as opposed to similar drinking behavior I'd witnessed in Europe and in the Americas, in the Reykjavik bar there was very little shouting. And, that night anyway, there was no fighting—simply heavy drinking.

The girls asked us where we were from. They mostly spoke English with an Icelandic accent, and every one of them wanted to go to America. When I mentioned that I was flying directly to Paris from Reykjavik, one of them, beautiful beyond reason, begged me to take her along. We danced.

Our foray into town turned out to be as rewarding as our first morning on the Laxa in Kjos, and early the next morning one of our newfound roommates drove us back to the lodge in her parents' car. Olive Oil did not look happy, so Russell invited her to go fishing, and from then on the tall, emaciated girl from the neighboring village followed us everywhere.

Once the fish moved into and up the river, everyone caught salmon. All the beats produced stories of fish that got away, fish of exceptional size, the flies they took, and those they refused. Weaving its way through each story was the magnificence of the Icelandic country and its weather that painted a hundred separate landscapes each day.

At the end of the week, when it came time to deal with Bergulfur and the sides of salmon everyone had ordered, all our friends were short on the exchange and wound up paying huge sums of money for slabs of fish they could have bought cheaper in New York. Russell and I caught ninety-eight salmon in five days, and on top of the fish we ordered, we were handed checks for the difference. Mine was for two dollars and eleven cents. Russell's was for almost five dollars. It was, apparently, the first time in the history and logbooks of the Laxa in Kjos that the anglers came out on top.

BILL

Bill Poppell was born in Coon Bottom (named because of its vast concentration of raccoons), east of Havana and a couple of miles west of the Ochlockonee River, in 1938. As a young man he fished the river to feed his family as much as for fun, returning at nightfall carrying a heavy croker sack full of catfish, bream, and bass. He used crank reels, a short stiff rod, and forty-pound test braided line for bass, and cane poles for the catfish and bream. His cooking method involved grease and heat. Hush puppies, black-eyed peas, okra, and biscuits accompanied the fish. Sweet tea and coffee followed. As opposed to the proliferation of southern Florida's high-rises, traffic, and congestion, Coon Bottom hasn't changed in fifty years, the exception being the health of the Ochlockonee River that Bill and his childhood buddies so loved to fish. A series of chemical spills over the years has reduced the numbers of fish by tenfold. As usual, little if any punishment has been conferred on the guilty.

One calm January afternoon Bill loaded cane poles, a couple of light spinning rods, and a glass jar full of minnows into his boat and pushed off the pond landing. His boat is larger and more stable than mine. Most days he takes his little springer spaniel with him. Today it was me. Once we were settled, he aimed the boat at the bank across from the landing where the speckled perch habitually make their beds.

Crappie (as they are known in the South), and all the other good eating fish in the pond except for catfish, belong to the sunfish family. The perch spawn about the full moon in January and are Bill's favorite to catch and eat. He is not alone. I know full-grown men across the South who would rather catch specks than watch football.

When I lived in southern Florida, Lake Okeechobee offered the best crappie fishing in the state, so it seemed natural to me that we should introduce them to the pond. However, I was told in no uncertain terms by fishing friends, even some pond experts such as Charles Mesing: "They love fish eggs, and they devour fry." "Specks multiply like rats." "They overrun ponds." And so on and so forth.

Bill quite rightly reasoned that we could keep the numbers of specks down by eating a mess of them every year. He released a dozen black crappies into the pond ten years ago, and now we both enjoy fine fishing, particularly in January, when the males protect the beds.

Speckle perch have a deep and laterally compressed body commonly seen in other panfish, such as bream. A rare fish might weigh four pounds, but the average crappie runs a pound; two pounds or more and it earns the moniker of "slab." The fish's fins are large and round and black. Its back is a dark shade of green fading to dirty yellow green (mostly apparent after it is caught) on an otherwise silver flank. Random black markings mottle its entire body and glow when the fish is lifted out of the water. The back of a crappie is arched, and its small head slopes down from its forehead past its eyes to a hawk-shaped nose. Its mouth is similar to a largemouth's in that its lower jaw extends to below the perch's eyes and juts past its top jaw. The fish is sometimes referred to as "paper mouth" because its lips are tender and will tear if undue pressure is applied to the hook.

The surface of the pond was winter dreary, the reflections of the pine trees diffused under the weight of the overcast. A north wind further confused the surface; the basso profundity of bullfrogs was but a distant memory. We fished just as I had fished for specks on Lake Okeechobee as a kid.

Bill likes to thread the small hook under the dorsal fin of the live bait. "It makes them swim cripple looking," he said as he handed me a long, telescopic fiberglass rod. With our minnows dangling in four feet of water, he allowed the wind to drift the boat over the beds. When my bobber dropped underwater, I followed his instructions: "Don't strike straight up in the air. Push the tip of the pole to one side, just like you were moving the minnow from one place to another."

It worked. The first fish of the morning pulled the bobber at an angle farther under the surface. With no reel and no drag to absorb the perch's struggles, the pole bowed and danced in my hands. When the fish tired, Bill was waiting with his net in the water. Moments later the black-and-white calico bass was in his cooler. We caught sixteen crappies on minnows that morning and six more trolling three-inch-long Rapala Diving crankbaits fashioned to resemble shad. When Bill fishes, he keeps all the fish he catches to eat, and the fish know it.

A month later, when the weather had warmed, Bill invited a group of his friends over for a fish fry in Coon Bottom, where he had lived surrounded by his siblings for seventy-six years.

A row of juniper and cedar trees shields his manicured lawn and double-wide from County Road 12, situated ten miles east of Havana. An organic garden sits between the outbuildings, ready to plant. Off to one side Bill had transformed a carport into a party room with wooden tables, metal chairs, a refrigerator, an oven, and a sink. "A fine place to feed a mess of people," his wife, Mary, exclaimed cheerfully.

On the north side of his four-acre compound there is a covered blind made of wood and topped by a corrugated roof. Bill sits on a fold-up chair next to his little springer spaniel during the hunting season and waits for doves to fly to the scratch feed he sows daily on the lawn. The blind is tall and narrow and purposely built to leave a restricted angle to shoot from.

"I made it so I can't pepper the road or my neighbor's house," he said. "As a result, no one bothers me."

In the summer of his eighth year, Bill went to work in the tobacco fields of Gadsden County, running bales of tobacco leaves from where they were cut to the mule wagons that carted them to the drying barns. He made fifty cents a day, two dollars and fifty cents a week. He has worked ever since.

When my wife and I arrived, Bill was cooking. He was surrounded by half a dozen of his cronies standing or sitting in metal chairs on the cement floor under the lean-to that extends out from the roof of one of the buildings he uses for storing what he has accumulated over the decades: metal plows disks, monkey wrenches, tongue-and-groove pliers, gimlets, drills, and more. Hanging off a low beam is an assortment of faded wooden fishing plugs Bill used as a kid. "Caught me plenty of bass with those before they turned to making them out of plastic," he said.

We all stood under a blue awning extending from his remodeled double-wide trailer. It was warm outside. Kids of all ages ran in and out of the house from the video screens indoors to the garden outdoors, where they chased each other like I used to chase my sister and her friends when I was young. Fifty of his guests were already there, and more were slowly making their way from their cars, parked in front of his dove blind. Mary greeted them. She was happy at the turnout.

The average age of the grown-ups hovered around seventy. Family, friends, and acquaintances from South Georgia and North Florida, men and women who had been working for the past sixty years inside and out twelve hours a day for nominal wages. Time and weather had changed their comportment from the kids they once were to the old folks they were now.

Four narrow Cajun fryers full of canola oil were heated to 350 degrees. Individual frying baskets filled with fish, french fries, and hush puppies were being manned by Bill and one of his friends, a tall dark-haired man who looked slightly put upon. The bass and catfish had been cleaned, scaled, and filleted. The bream and specks had lost their heads to a diagonal decapitation. All species and size of fish had been sorted out, separated, and placed on metal cookie sheets ready to be rolled in Aunt Jemima cornmeal mix before being dropped into the hot oil.

Gordon, a short man with a full head of gray hair and Bill's oldest friend, poured me a glass of sweet ice tea, and when I asked how he had been (I hadn't seen Gordon in a year), he grinned: "I'm kicking right on."

He knew I liked to hear about his childhood as Bill's friend and their adventures together. Soon he said, "One time, Bill and I were paddling down the Ochlockonee in a canoe, with me in the bow, and every time I spotted a moccasin on a tree limb, I knocked it into the boat!" They both laughed.

It sounded a lot like some of the asshole things Gil or I used to do to each other half a century ago. Then it occurred to me that we were of the same generation.

Gordon turned to a very old man sitting in a lawn chair drinking coffee: "At least in our time, we knew a little about everything. The kids now days don't know nothing about anything!"

As is the custom in the South, before any of the food was touched, grace was said by a pastor friend of the family to a gathering of bowed heads: "O Gracious God, we give you thanks for your overflowing generosity to us. Thank you for the blessings of the food we eat and especially for this feast today. Thank you for our home and family and friends, especially for the presence of those gathered here. We ask your blessing through Christ your son. Amen."

And the supper was on.

The fries and hush puppies had been frozen, and tasted like it, but the fish, caught in the pond and the Ochlockonee River, tasted clean and fresh like the water they had lived in. Bill and his tall friend understood heat, and the degree of browning cornmeal would allow in order to keep the meat firm and flaky under a crunchy coating.

Casseroles of cheese grits and others of noodles, black-eyed peas, lima beans, potato salad, and buns lined the long wooden serving table. One after another we took turns filling large paper plates individually molded to separate the ingredients. Dessert cakes and pie rested on a separate table next to the door. In order to find room for a fish on my plate, by the time I reached the platter, I had no choice but to lay the perch on top of the mountain of food I had selected walking down the line. A heavyset lady nodded at my plate and turned back to the serving table. "I might could have me another helping," she said smiling.

My fish was as well cooked as any celebrity chef in New York or Las Vegas could ever wish for.

When we left the party, Gordon raised his hand and said, "Good to see you. Give me a holler. We'll go fishing."

RAIN

In February, a few weeks after Bill's fish fry, the Weather Channel predicted that the front moving in from the west would drop a significant amount of rain on Tallahassee. It had been three years since the beginning of the drought, six months since a storm had lit up the farm. The first rumblings of thunder sounded like someone walking on bare heels on the roof above my head. Outside the pond house it was dark and still. Moments later I heard the sound again, this time followed by a finger of light that momentarily brightened the window above the sink. It was chased by uncommon winter grumblings in the west. After that, nothing. I assumed that just as it had happened in the past, the weather streamed north of the farm before reaching it.

But the first squall skittered across the pond in the middle of the night. By dawn a constant downpour obscured the dock. The wind blew, and the clouds loosened what is referred to in the South as a "toad strangler." Over the next four days the storm dropped fourteen inches of rain on Gadsden County, and except for the first night, when the window brightened over the sink, I never saw any lightning. What thunder I heard was always a long way off, but that didn't deter the rain.

It came down flat, hard, and determined.

On the first day, to ease the profound thirst the drought had burned into the ground, the earth captured all the fallen water,

just as it had twenty years earlier. Standing on the porch watching the rain, I felt the earth breathe a sigh of relief.

On the second day the ground, now satiated, assisted the
rain down the slopes and into the pond, the runoff measurable.
The sight of running water has been rare on the farm; the flow
sounded like a shallow brook. I cupped my hands and drank.
By the third day the water had shaped its preferred routes in the
grass, and cascades of water flowed through them. When the
rain fell at its hardest, the drops bounced straight up off the surface. Up close they acted like small balls of mercury dancing in
harmony. When the rain paused, fog rose from the water, until
the next downpour flattened the mist and erased the ambiguous
shadows of the pine trees.

At one o'clock in the afternoon on the fourth day, the rain
stopped abruptly. Three hours later the sun was out and all was
quiet again. The small clearings in the forest of pine trees on
the hill across from the cabin were open to the sky, the dappled
light bright and gray. Sunlight climbed up the trunks of trees,
and the shadows of loblollies once again stretched across the
surface of the pond.

From the chair where I had spent the last three years watching the pond's stature diminish, its banks grow wider, and tall
grasses rise doggedly in its shallows, the surface level of the
water was suddenly higher and therefore much closer than it
had been. This proximity forced my gaze upward.

The trees we had pushed into the water the year before to
provide fish with a new campus were now covered with an additional two and a half feet of water. The bream born during the
summer sparkled like snowflakes inside the safety of the limbs.
The bass fry hiding among the partially submerged trees moved
closer to shore and into the acres of newly created grass cover.
In the spring that followed, the bass fry would be joined by shad
fry and then by crappie fry and then once again by bream fry.

After the rain a thousand robins investigated the resurgence of worms and mole crickets on the sodden earth. Hibernating snakes, woken and wet, collected themselves in the sun at the base of hollow tree stumps to dry. Peepers sang their enthusiasm. Hungry sparrows fluttered from the branches of the oak tree to the feeder below.

All the cover and potential food hiding in this new ecosystem agitated the bass and encouraged their desire to hunt. At the head of every dead tree, now five feet underwater, a big female bass stood guard over her domain.

On shore spring was a month away. There will be burning, plowing, and walks through harrowed fields in search of artifacts. The dogwoods will bloom, and the bluebirds will take to the boxes we have made for them to nest in. Fox squirrels will chase each other up the loblolly pines, dislodging slivers of bark the size of sparrow wings. In the summer I will hear the bobwhites sing and watch flocks of vesper swallows slither to the pond and, after a dip, shiver their feathers free of water. I'll see the cotton-colored mouths of moccasins coiled around the knees of cypress trees and watch swallow-tailed kites swoop down from the thermals to gather grasshoppers. In November the red-tailed hawks will hunt field rats discernible from above inside the thinning cover, eagles will elegantly confiscate catfish in front of the cabin, and the cypress trees surrounding the pond will once again light up and cast golden impressions impossible to ignore.

When I think of those four days of rain that refilled the pond, I think of the future evolution of nature, the struggle for life in varying degrees of desperation. I think of the sun and the influence of water on life, the patience of books, and the timbre of songs, and I envy the fish and the birds, empowered to fly free and swim without terrestrial barriers.

I think of my grandchildren, who are made of music, and I want them to dance.

EPILOGUE

Tonight the cabin is a sad place for me. A loner by choice, I have returned again to my solitude, a state of being that has pursued me all my life. Questions arise after dark, questions I cannot answer. Questions that speak to the meaning of life.

A few years ago I noticed the first liver spot on the back of my hand. The blemish did not wash off. More appeared. Now my hands are evidence of more than half a century of wind and sun. The thin skin on my fingertips reminds me of my father's. As a child I pretended to sleep when I felt the caress of his fingers on my cheek. Thirty years later, when he was old, my father would wait for months for my visit from across the Atlantic, a trip I did not make often enough to ease my guilt.

Similarly, my old dog waits hours on end for the slightest token of affection or a small sign of recognition, both of which she received in abundance when she was young. Now I help her climb up on things she could jump up on when I started this book, and that too breaks my heart.

Since the alternative of mingling with my peers is unpleasant, I wonder if the company of dogs, birds, and fish will suffice as I grow even older. I wonder if nights like these will impose denial, as they did to my father, on the assurances of a two-thousand-year-old religion he questioned to the end.

The miracle of nature has sustained me for seventy years. I believe in silence and in rain and in the innocence of children, and I have faith that every blade of grass, every drop of rain, polluted or not, will flourish again after we are gone. Meanwhile I face the sun and tow my shadow. And when everything is said and done, I'll move on to whatever is meant to be, which I suspect is nothing.

Remembering the details of one's life decades after the fact is not a simple matter. It is all too easy to be sidetracked by memories of specific incidents and feelings of melancholy that have nothing to do with the reminiscences at hand. Real and imagined wounds are reopened. Mundane events assume an importance they do not deserve. The death of family and friends and pets are more deeply mourned, and the heartfelt pain of loss is ever present.

Something rather awful has influenced my body and softened my spirit. The riot that lived in my soul as a child has been translated into the unvoiced but insistent rebellion of a sentient being with an eye on the finish line.

It is tomorrow already and yesterday is every day.

The past drags at me like soft marl, chiding me for growing old and forgetting what it felt like to be someone else's child. I am father and grandfather now, and what little of that island boy is left detests his image in the mirror.

For a minute, a week, a year, perhaps a decade, I was the long-legged heron of the flats. Barefoot and naked, I hunted for shallow-feeding fish in ankle-deep water, casting at every shadow, including those of passing clouds. The heron is not dead, but he has aged into a solitaire, shunning the company of others and troubled at the future he will bequeath his offspring.

I dispute the common belief that life is short. For anyone past the age of sixty, life is a long experience, in my case filled with the beauty and unpretentiousness of the natural world with which I have surrounded myself. When I fall asleep in the cabin next to the pond, memories surface in the form of dreams that mysteriously bend time and summon whatever haunts the alcoves of my mind.

The impenetrable magic of a single dream turns the clock back half a century just as effortlessly as it conjures yesterday.

ABOUT THE AUTHOR

Guy de la Valdène was born and raised in France. His earlier books include *For a Handful of Feathers*, *Making Game: An Essay on Woodcock*, *Red Stag*, and *The Fragrance of Grass*. His articles have appeared in *Gray's Sporting Journal*, *Sports Afield*, *Garden & Gun*, and *Field & Stream*, among other publications. He lives on an eight-hundred-acre farm outside of Tallahassee, Florida.